Benjamin Mortimer Hall

**A Preliminary Report on a Part of the Waterpowers of Georgia**

Benjamin Mortimer Hall

**A Preliminary Report on a Part of the Waterpowers of Georgia**

ISBN/EAN: 9783744662048

Printed in Europe, USA, Canada, Australia, Japan

Cover: Foto ©berggeist007 / pixelio.de

More available books at **www.hansebooks.com**

*With the Compliments of*

*W. S. Yeates,*

*State Geologist.*

WATER-POWERS OF GEORGIA  FRONTISPIECE.  PLATE 1

THE WITCH'S HEAD, TALLULAH FALLS, GEORGIA.

GEOLOGICAL SURVEY OF GEORGIA

W. S. YEATES, State Geologist

BULLETIN No. 3—A

# A PRELIMINARY REPORT

ON A PART OF THE

# Water-powers of Georgia

COMPILED FROM THE NOTES OF

C. C. ANDERSON

Late Assistant Geologist

AND FROM OTHER SOURCES

BY

B. M. HALL, Special Assistant

1896

GEO. W. HARRISON, STATE PRINTER
FRANKLIN PRINTING AND PUBLISHING COMPANY
Atlanta, Ga.

# ERRATA

On page 125, 5th column of table, in fourth line from bottom, for "— 1.55," read — 0.55.

On page 128, foot-note at bottom, for "inches," read *feet*.

# THE ADVISORY BOARD

## of the Geological Survey of Georgia

(Ex-Officio)

HIS EXCELLENCY, W. Y. ATKINSON, Governor of Georgia,

PRESIDENT OF THE BOARD

HON. R. T. NESBITT . . . . Commissioner of Agriculture

HON. G. R. GLENN . . . Commissioner of Public Schools

HON. R. U. HARDEMAN . . . . . . . State Treasurer

HON. W. A. WRIGHT . . . . . . . Comptroller-General

HON. A. D CANDLER . . . . . . . . Secretary of State

HON. J. M. TERRELL . . . . . . . . Attorney-General

SEPTEMBER 5TH, 1896.

*To His Excellency,* W. Y. ATKINSON, *Governor, and President of the Advisory Board of the Geological Survey of Georgia,*

SIR: — I have the honor to transmit, herewith, a preliminary report on a part of the Water-powers of Georgia, compiled by Mr. B. M. Hall, Special Assistant, from the report of Mr. C. C. Anderson, late Assistant Geologist, and from other sources.

During the past year, especially, there has been great demand for information, as to Southern water-powers, coming mostly from Eastern manufacturers, contemplating the establishment of cotton-mills and other factories in the South. This is, therefore, an opportune time, for the issuing of such a bulletin, which will be the first of a series on this subject. Field-work, for a second bulletin, is now in progress; and, as soon as sufficient data has been collected, a second report will be submitted.

Very respectfully yours,
S. YEATES,
State Geologist.

# WATER-POWERS OF GEORGIA

## CHAPTER I

### INTRODUCTORY[1]

The necessity for an economic survey of the water-powers of Georgia, that would show their number and their degree of availability for practical use, has long been felt. So, when the office of State Geologist was revived, in the fall of 1889, by act of the legislature, with an appropriation for five years, from July 1st, 1890, Dr. J. W. Spencer having been elected State Geologist, a survey of the water-powers of the State was begun, by Mr. C. C. Anderson, Assistant Geologist, who continued field-work during field-seasons, until the close of the season of 1892. During this time, Mr. Anderson established gauge-stations and appointed gauge-readers, who made regular monthly reports of daily readings, for a period of thirteen consecutive months. These stations were established at certain points along the Chattahoochee, Flint and Ocmulgee rivers, and some of their tributaries. Mr. Anderson proceeded to make surveys of the shoals, by soundings along cross-sections, and by measuring the velocities of the streams, with a Haskell current-meter. He was under instruction, too, to make certain geological and timber observations, and to collect

---

[1] By W. S. Yeates, State Geologist.

specimens of minerals, rocks and soils, all of which he did. His report was submitted to the Geological Board, about the time the Survey was reorganized in April, 1893; but it was never published. In the course of a year's experience, it became apparent to the present State Geologist, that a published report on the water-powers of Georgia was greatly needed; for many inquiries for information on this subject were constantly coming to him from manufacturers and others outside the State. By advice of the Geological Board, a competent hydrographic engineer was employed, to carefully examine Mr. Anderson's report, and, subsequently, to compile, from it and other reliable sources, material for a preliminary bulletin on such of the water-powers in the State, as had been surveyed. Such published work as was done by the Survey, when Dr. George Little was State Geologist, from 1874 to 1879, has been made use of; and the United States Weather and Census Reports have, in a measure, contributed to this bulletin. While it cannot be claimed, that this report is complete, even as to the rivers and tributaries undertaken; yet, it will serve to call attention, in a practical way, to a large number of valuable water-powers, by far the greater number of which are unutilized. On the map, which accompanies the report, there are a few omissions, which were occasioned by the compiler's failure to submit data, before the engraving was completed and the transfers were made. Along the tributaries of the three principal rivers, a number of water-powers are not given, because no surveys have been made. These will be surveyed, and included in the next one of this series. Arrangements have been made, by which this Survey is now working conjointly with the U. S. Geological Survey on the Water-powers of Georgia. This plan of co-operation gives to each Survey the data collected in the field by the other, whereby each is enabled to cover more territory, in a given time, than it otherwise would be able to do.

PLATE II

FALLS OF L'EAU D'OR, TALLULAH FALLS, GEORGIA.

## CHAPTER II

### THE RECENT INCREASE IN THE VALUE OF WATER-POWERS, ESPECIALLY THOSE OF GEORGIA

Very few of the large water-powers of Georgia are utilized. This is a fact, not from lack of energy and enterprise in the people of the State; but because their energy has, heretofore, been directed mainly to agriculture and commerce, and not to manufacturing. But a rapid change is taking place in this respect; and it is all the better for our future, that this, the dawn of the age of electricity, has found us with undeveloped powers, ready to receive the latest and best machinery, without the loss and expense of taking out old machinery, to make room for it; or, worse still, the necessity of running the antiquated machinery at a great loss, when it is brought into competition with the latest improvements.

This bulletin locates, and gives some information concerning, hundreds of water-powers in the State, many of the smaller being utilized, and a few of the larger, partly utilized; but by far the greater number are absolutely in their natural state.

The following are some of the great powers, in the State, that are running to waste:—

*Tallulah Falls*, in Rabun county, with a 335-foot fall.

*Coosawattee Shoals*, in Gilmer and Gordon counties, a succession of cascades for seventeen miles.

*The Etowah Mining Co's. Shoals*, at Cartersville, on the Etowah river, with a fall of 50 feet.

*The Great Amicalola Shoals*, in Dawson county, with a 234-foot fall.

*Roswell and Bull Sluice Shoal*, on the Chattahoochee river, in Fulton county, fourteen miles from Atlanta, with 50 feet of fall.

*The Vining Shoals*, on the Chattahoochee river, in Fulton county, seven miles from Atlanta, with a fall of 32 feet.

*The Jack Todd Shoal*, on the Chattahoochee river, in Harris county, near West Point, with 51 feet of fall.

*Hargett Island Shoals*, on the Chattahoochee river, in Harris county, with 60 feet of fall.

*The Great Shoals*, on the Chattahoochee river, at Columbus, with 120 feet of fall.

*Flat Shoals*, on the Flint river, in Pike and Meriwether counties, with 32 feet of fall.

*Yellow Jacket Shoals*, on the Flint river, in Upson county, with a 36-foot fall.

*Rogers' Shoal and Nelson's Shoal*, on Big Potato creek, Upson county, with 81 feet and 115 feet of fall, respectively.

*High Falls*, on the Towaliga river, Monroe county, with a fall of 96 feet.

*Sweet-water Shoals*, on Sweet-water creek, Douglas county, near Austell, with an 80-foot fall.

*Cedar Shoals*, on the Yellow river, in Newton county, with 55 feet of fall.

*Garner Shoals*, on Alcovy river, in Newton county, with a fall of 85 feet.

*The Harper or Pittman Shoal*, on the Ocmulgee river, in Butts county, with a 28-foot fall and a six-foot shoal just below it.

*Tallassee Bridge Shoal*, on Middle Oconee river, in Jackson county, with a 52-foot fall.

*High Shoals*, on the Apalachee river, in Oconee county, with 50 feet of fall.

*Barnett's Shoal*, on the Oconee river, in Oconee county, with a 54-foot fall.

*Trotter's Shoal*, on the Savannah river, in Elbert county, with 75 feet of fall.

*Anthony Shoal*, on the Broad river, in Elbert and Lincoln counties, with over 70 feet of fall.

These powers are mentioned here, to attract attention to the tabulated statements of Chapter III, where they, with numerous others, are given in detail.

Water-power has always been recognized as the cheapest and best power for running stationary machinery. Hence, in all manufacturing countries, the powers, that are conveniently located, with reference to transportation, and capable of being developed at a reasonable cost, have formed the nucleus for important industrial towns. As these towns have grown, and offered advantages for manufacturing, beyond the capacity of the available water-power, steam-power has been added, rather than go to other and less favorable localities, for more water-power. This is why such cities as Lowell, Mass., use more steam-power than water-power — a fact that has furnished a pretext for all kinds of unreasonable arguments, to prove that steam-power is cheaper than water-power; arguments, that are made by people interested in the manufacture of steam-engines, the development of coal mines, or the prosperity of towns, not blessed with water-power. It has been freely admitted, by all advocates of water-power, that it is often cheaper to erect and operate a steam-plant in a favorable locality, than to develop and run a water-power, where there are no facilities for transportation; and this fact has caused many fine water-powers to remain undeveloped. But the recent improvements, in electric motors and long distance transmission, have brought about a new era in water-power development. As factories could not go to these water-powers, the water-powers are beginning to come to the factories; and, not only to the factories, but to the operation of railroads, a field which has, until recently, been considered the exclusive domain of the steam locomotive. It does not even stop at this point; for it is rapidly displacing coal-gas and steam-generated electricity, in lighting our cities; and it may soon perform an important part in cooking and heating.

The old idea of development was to bring a power-canal into a city, and build factories along the canal; but many cities, located on or near rivers, having fine shoals, are prevented from doing this, by topographical difficulties, that are practically insurmountable. With the possible exception of Macon and Milledgeville, the only city in Georgia, favorably located for this kind of power-development, is Augusta; and it is highly probable, that the Augusta power-canal, constructed in 1847, is the only one of the kind, that any Georgia city will ever possess. There is no longer the same necessity for this kind of development. The modern plan of placing a generating-plant at the shoals, and transmitting the power, electrically, for distribution wherever it is needed, is, in most cases, infinitely better; and the day is not far distant, when many towns, situated in or near the Crystalline Belt of Georgia, can have all the power desired, at a much smaller cost than steam-power. Capitalists are now contemplating the taking hold of an enterprise to develop the large powers on the Chattahoochee river, near Atlanta, for this purpose; and other cities in the State are also planning to make use of contiguous water-powers, in the same way.

The foregoing discussion is to show the great possibilities for water-power, as a source of city-power, and its corresponding increase in value. It is not intended to intimate, that the powers of this State are less conveniently located for factory-sites, than those of other States. On the contrary, many of the best water-powers are close to important railroads, and offer beautiful locations for manufacturing towns. Many others, near railroads, but situated in deep gorges and among rock-cliffs, can be profitably utilized, by placing a power-station at the shoal, and transmitting the power, electrically, to a good factory-site on the railroad.

There are also many valuable powers in our mining and quarrying regions, that can be utilized in like manner. The granite quarries of Lithonia and Stone Mountain can be run by power from South river, near at hand. The marble quarries of Long Swamp valley in Pickens county, where more than two million dollars is already invested in developments, can be run by power from the Amicalola river, eight miles distant; and the gold mines, that cover a large area in the State, can have cheap power from the adjacent streams, for running drills, ventilators and hoisting and milling machinery, thus encouraging deep mining, which is so necessary to the proper development of such properties. It is now an acknowledged fact, that cotton-goods can be manufactured more cheaply in the South, than anywhere else; and the bringing of the cotton-factories to the cotton-fields, which has already been begun in earnest, will continue, until the greater portion of our cotton crop will be shipped in the form of manufactured goods. Eastern capitalists, seeing and acknowledging this tendency, are beginning to investigate our region, with a real desire to find out something about it.

It is expected, that the Cotton States and International Exposition, recently held in Atlanta, will largely increase the demand for information along this line; and this bulletin, the first of a series on this subject, is compiled for the purpose of giving such information, concerning our water-powers, as is attainable from the data, thus far collected.

# CHAPTER III[1]

## THE STREAMS AND DRAINAGE BASINS OF GEORGIA, WITH TABLES SHOWING TRIBUTARIES AND WATER-POWERS

### DRAINAGE BASINS

A study of the water-courses of Georgia is peculiarly interesting. The streams all rise within the borders of the State, and flow to the four points of the compass, forming a large number of separate and distinct drainage basins, which discharge into either the Gulf of Mexico or the Atlantic Ocean, at points very remote from each other. The nine principal drainage basins,[2] that lie wholly or partly in the State, are:—

FIRST — *The Tennessee Basin*, occupied by tributaries of the Tennessee river, whose waters find their way through the Mississippi to the Gulf, below New Orleans.

SECOND — *The Mobile Basin*, in which originate the Coosa and Tallapoosa rivers, with their outlets into the Gulf at Mobile.

THIRD — *The Apalachicola Basin*, through which run the waters of the Chattahoochee and the Flint rivers, reaching the Gulf at Apalachicola.

FOURTH — *The Altamaha Basin*, including the Oconee and Ocmul-

---

[1] By authority of the Geological Board of Georgia, this chapter was furnished by the State Geologist to the Commissioner of Agriculture, for use in "Georgia: Her Resources, etc.", published in 1895.

[2] See map, page 16.

(14)

gee waters, which enter the Atlantic Ocean, by way of the Altamaha river.

FIFTH — *The Ogeechee Basin*, which is drained into the Atlantic Ocean, by the Ogeechee river.

SIXTH — *The Savannah Basin*, which is drained by the Savannah river into the Atlantic Ocean.

SEVENTH — *The Ocklockonee Basin*, which is drained into the Gulf through Ocklockonee bay.

EIGHTH — *The Suwannee Basin*, which is drained into the Gulf by the Suwannee river.

NINTH — *The Satilla and St. Mary's Basin*, the rivers of which flow into the Atlantic Ocean near Cumberland Island.

Five of these basins, the Tennessee, the Mobile, the Apalachicola, the Altamaha and the Savannah, have a great portion of their territory lying in the Crystalline Belt of the State, which is all that part of the State north of a line joining Augusta, Macon and Columbus, and east of a line passing through Polk, Bartow, Gordon and Murray counties. These two lines are shown on the map, and are designated, respectively, as the Southern Fall Line and the Western Fall Line. It may be said, in a general way, that the greatest water-powers in the State are at, or not far above, the points where the rivers cross these fall lines; but it must not be understood from this statement, that the greater part of the total water-power of the State is in the vicinity of these fall lines. These streams are a series of shoals from their heads to the fall line, which is the head of navigation in all the rivers, except the Etowah, and which marks the divide between the hard granite and schistose rocks of the older Crystalline region and the softer materials of a younger formation; but the last great plunge that the river makes, in its descent, forms a water-power, that is more important, than any other along its course. To illustrate: — The Chattahoochee river, from Thompson's bridge, in

Hall county, to West Point, a distance of about 180 miles, falls 389 feet, while from West Point to Columbus, a distance of only 34 miles, it falls 362 feet.[1] About 120 feet of this is in the last four miles above navigable water. There is no other four-mile section of the river, that has so great a fall. It is thus seen, that, while the river has a very large amount of available power along its upper course, the combination, at Columbus, of great fall and great volume makes a most valuable water-power, the largest in the State, being nearly 80,000 gross horse-power at average low-water. It is also true of the Oconee, Savannah, Ocmulgee, Etowah and Coosawattee, that they have a greater concentration of power at or near the limit of the Crystalline rocks, than at any other single point; but the rivers of the Atlantic slope occupy lower basins in the Crystalline region, than that of the Chattahoochee, while the Paleozoic country, immediately west of the Western Fall Line, is much higher than the Tertiary region, south of the Southern Fall Line; consequently, these rivers have no such shoals at the fall line, as those on the Chattahoochee at Columbus.

A striking characteristic of the Savannah and Ocmulgee rivers is the great height of the shoals on their large tributaries; notably, Tallulah Falls and Anthony Falls of the Savannah basin; and the high falls on the Towaliga, Alcovy, Yellow and South rivers of the Ocmulgee water-shed.

It will be readily understood, from the foregoing, that the important water-powers of the State are confined mainly to the Crystalline region, where the fall is steep, and the country-rock is gneiss and micaceous slates. These streams drain off most of the rainfall, that is not evaporated. Being in a region, where the rainfall is remarkably uniform throughout the year, they can be relied on, for constancy of supply.

---

[1] The river cuts through Pine Mountain Range (the Gulf Coast Range), about half way from West Point to Columbus.

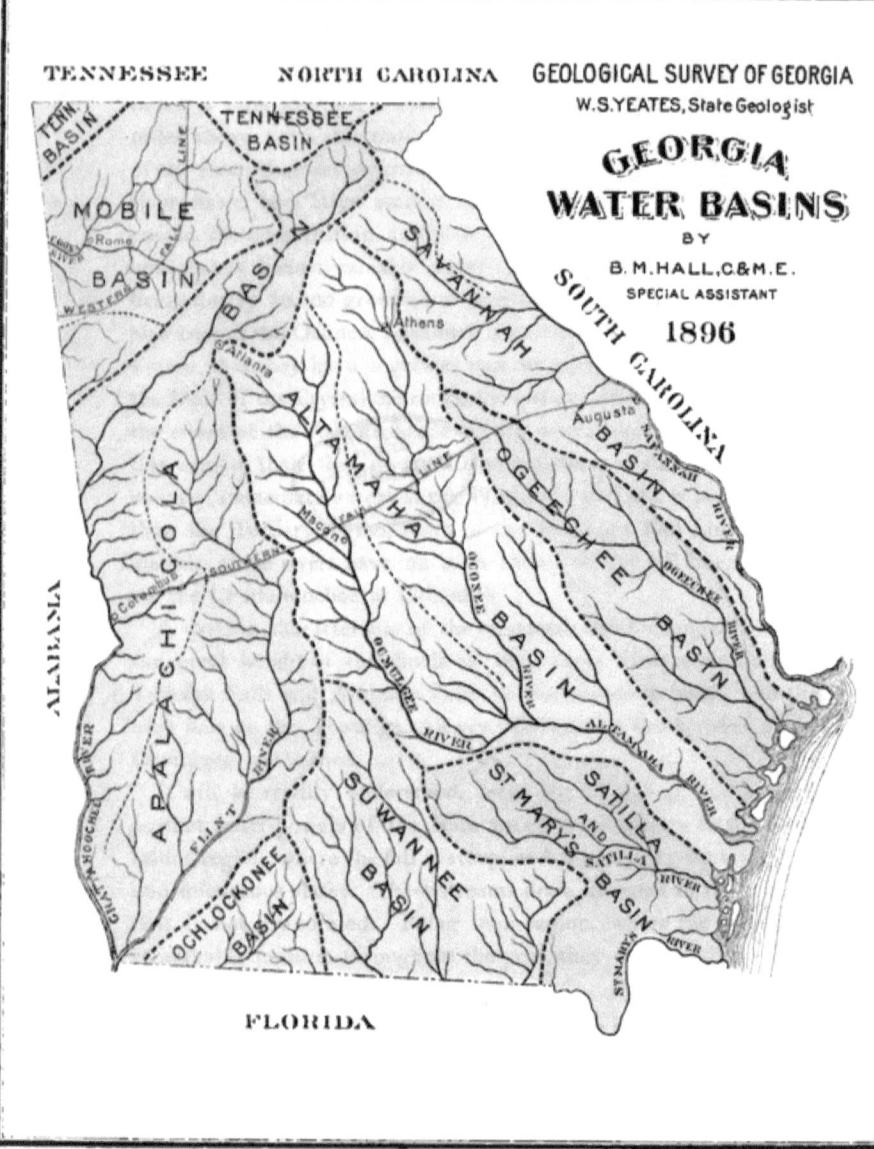

Special attention may be called to the form and position of the Chattahoochee water-shed. It is very narrow in proportion to its length and depth. Its greatest breadth is in the Blue Ridge mountains of Lumpkin, White and Habersham counties, where the autumn rainfall is nearly twice as great, as it is at Atlanta. The Atlanta rainfall may be taken as an average, for all that part of the Crystalline region, which is not mountainous. The table on the following page, showing this precipitation for twenty-six years, is from the records of the U. S. Weather Bureau.

## AVERAGE RAINFALL

### U. S. DEPARTMENT OF AGRICULTURE, WEATHER BUREAU
### STATION: — ATLANTA, GEORGIA

Data: — *Monthly Rainfall (Inches)*

| Year | Jan. | Feb. | March | April | May | June | July | Aug. | Sept. | Oct. | Nov. | Dec. | Annual | Observer |
|---|---|---|---|---|---|---|---|---|---|---|---|---|---|---|
| 1870 | 2.03 | 6.20 | 6.11 | 5.20 | 7.77 | 5.97 | 2.25 | 2.09 | 9.40 | 0.67 | 5.42 | 3.74 | . . . . | Maj. S. B. Wright, Ga. Agrl. Reports. |
| 1871 | 2.94 | 5.28 | 7.66 | 3.09 | 3.75 | 1.82 | 1.12 | 6.49 | 4.44 | 2.09 | 3.41 | 3.36 | 54.19 | " |
| 1872 | 2.94 | 5.28 | 7.66 | 3.09 | 3.75 | 1.82 | 3.91 | 5.84 | 2.26 | 0.74 | 2.12 | 4.48 | 43.89 | " |
| 1873 | 3.36 | 12.04 | 2.58 | 1.96 | 6.05 | 6.86 | 3.87 | 2.08 | 5.40 | 1.23 | 3.15 | 2.41 | 50.99 | " |
| 1874 | 3.14 | 6.86 | 7.38 | 10.42 | 3.00 | 7.71 | 4.70 | 10.00 | 0.47 | 0.80 | 3.19 | 3.00 | 60.67 | " |
| 1875 | 5.60 | 6.92 | 10.27 | 4.79 | 1.84 | 4.58 | 3.84 | 3.42 | 4.64 | 1.50 | 3.45 | 6.14 | 56.99 | R. J. Redding of Ga. Agrl. Dept. |
| 1876 | 3.32 | 5.37 | 5.59 | 6.01 | 5.00 | 3.25 | 3.49 | 5.32 | 0.82 | 1.81 | 2.56 | 4.35 | 46.87 | " |
| 1877 | 5.93 | 3.10 | 7.40 | 6.43 | 0.72 | 5.71 | 3.40 | 0.86 | 2.85 | 3.78 | 3.85 | 4.11 | 48.20 | " |
| 1878 | 3.76 | 6.54 | 6.72 | 3.15 | 2.25 | 5.39 | 1.77 | 3.76 | 1.75 | 1.99 | 4.54 | 5.80 | 47.42 | R. J. Redding of Ga. Ag. Dept. and U.S.W.B. |
| 1879 | 4.29 | 3.09 | 2.49 | 3.98 | 4.16 | 3.20 | 5.75 | 4.76 | 1.43 | 5.44 | 3.88 | 7.86 | 50.33 | U. S. Weather Bureau. |
| 1880 | 2.86 | 3.11 | 11.87 | 7.07 | 4.52 | 3.57 | 3.16 | 3.61 | 6.21 | 2.81 | 8.21 | 5.70 | 62.70 | " |
| 1881 | 8.35 | 10.41 | 10.08 | 4.58 | 1.27 | 2.46 | 0.56 | 4.10 | 3.76 | 3.44 | 4.30 | 7.53 | 59.74 | " |
| 1882 | 6.40 | 10.29 | 4.16 | 5.21 | 3.02 | 3.22 | 6.61 | 5.86 | 3.51 | 1.35 | 4.22 | 4.37 | 58.22 | " |
| 1883 | 15.82 | 3.22 | 3.73 | 8.20 | 2.00 | 3.31 | 1.06 | 2.73 | 1.38 | 1.52 | 4.72 | 4.84 | 51.53 | " |
| 1884 | 5.20 | 5.84 | 9.70 | 5.86 | 1.33 | 10.73 | 2.42 | 2.06 | 0.68 | 0.70 | 2.84 | 6.09 | 52.85 | " |
| 1885 | 8.44 | 4.14 | 4.26 | 1.31 | 6.12 | 4.83 | 4.02 | 6.92 | 6.51 | 3.04 | 3.08 | 2.64 | 57.01 | " |
| 1886 | 7.33 | 1.53 | 11.16 | 2.53 | 6.21 | 8.68 | 2.08 | 2.36 | 0.53 | 0.03 | 5.32 | 3.03 | 50.78 | " |
| 1887 | 3.52 | 3.74 | 1.09 | 1.38 | 1.76 | 2.82 | 14.11 | 7.51 | 4.20 | 3.28 | 0.30 | 5.79 | 50.40 | " |
| 1888 | 3.89 | 5.91 | 8.16 | 1.34 | 6.86 | 4.71 | 1.85 | 3.89 | 14.26 | 3.99 | 4.70 | 5.42 | 64.98 | " |
| 1889 | 6.39 | 5.28 | 2.49 | 2.54 | 3.16 | 5.03 | 8.83 | 6.73 | 6.32 | 2.21 | 5.17 | 0.60 | 54.75 | " |
| 1890 | 2.95 | 3.36 | 3.13 | 2.04 | 6.32 | 1.12 | 5.37 | 3.99 | 5.36 | 4.89 | 0.18 | 3.89 | 42.60 | " |
| 1891 | 6.73 | 8.50 | 10.16 | 1.58 | 2.17 | 4.71 | 5.38 | 2.59 | 1.19 | 0.02 | 3.26 | 3.68 | 49.97 | " |
| 1892 | 8.93 | 3.44 | 5.71 | 4.75 | 1.37 | 4.65 | 3.77 | 6.66 | 2.70 | 0.59 | 4.41 | 2.89 | 49.87 | " |
| 1893 | 3.02 | 5.45 | 2.43 | 2.48 | 4.46 | 4.65 | 2.13 | 4.07 | 3.06 | 0.39 | 1.11 | 3.18 | 36.43 | " |
| 1894 | 5.09 | 4.08 | 2.99 | 3.06 | 1.49 | 1.29 | 5.55 | 3.70 | 5.78 | 2.62 | 0.92 | 3.45 | 40.92 | " |
| 1895 | 5.47 | 2.01 | 7.55 | 5.20 | 3.99 | 4.87 | 2.75 | 8.55 | 0.21 | 1.30 | 1.04 | 2.08 | 45.92 | " |
| 1896 | 3.12 | 3.04 | 3.29 | 0.58 | 1.95 | 2.66 | 7.55 | 1.97 | 1.36 | 1.28 | 5.90 | 1.42 | 34.12 | " |
| Av'rge 26 y'rs | 5.30 | 5.37 | 6.16 | 4.03 | 3.56 | 4.53 | 4.19 | 4.61 | 3.48 | 2.07 | 3.47 | 4.19 | 50.96 | 1871 to 1896, inclusive. |

This Average Rainfall is distributed as follows: — Spring, 13.75; Summer, 13.33; Autumn, 9.02; Winter, 14.26.

## STREAMS

The following lists of important streams, and the accompanying water-power tables, give some idea of the extent and distribution of the water-powers of the State, and the work to be done, in order to arrive at a full knowledge of them. The tables are a compilation of data, derived from all available sources. In all the streams, covered by the surveys of Mr. Anderson, the compiler has computed, from his data, volumes corresponding to the lowest stages noted in his fluctuation-tables. While it is not claimed, that the low-water volumes, thus deduced, are absolutely correct, they are given as a close approximation of the true volumes, that would have been found by measurement, at the lowest stages noted in the tables. The data, as to other streams, has been derived from Janes's Hand-book of Georgia, Henderson's Commonwealth of Georgia, the 10th Census Report of the United States, and from other sources. It is mostly of a general nature, serving to call attention to certain water-powers, without giving definite information concerning them. The fall, where given, is probably accurate, as the surveys were made by engineers of high standing; but the measurements of volume, though correct for the time they were taken, give little information, as to the flow of the stream throughout the year. The volumes, given by the U. S. Census Reports, are estimated from the area of water-shed, and are used, only, when there is no positive information obtainable. The tables of utilized power are from the 10th U. S. Census Report, being the only data at hand. Mr. Anderson's statistics of utilized power are given in the regular power-tables; but they cover only a limited area.

In these tables, the column, "Source of Information" shows the work of the Georgia Survey, 1874-79, by the names, C. A. Locke and D. C. Barrow. The names of Messrs. Frobel, Sublett and Carson represent surveys by the U. S. Government.

## TENNESSEE BASIN—IMPORTANT STREAMS

| STREAM | TRIBUTARY TO | COUNTY | REMARKS |
|---|---|---|---|
| Nickajack Creek | Tennessee River | Dade | |
| Lookout Creek | " | " | |
| Chattanooga Creek | " | Walker | |
| Chickamauga River | " | " | |
| West Chickamauga Cr. | Chickamauga River | " | |
| Middle Chickamauga Cr. | " | Catoosa | |
| East Chickamauga Cr. | " | Whitfield | |
| Toccoa River | | Fannin | |
| Fightingtown Creek | Hiawassee River | " | |
| Hemptown Creek | Toccoa River | " | The streams of Fannin, Union and Towns Counties are a succession of shoals, from their heads to the State line; but no surveys have been made of the water-powers. |
| Nuntootlee Creek | " | " | |
| Notteley River | Hiawassee River | Union | |
| Cooper's Creek | Tennessee River | Towns | |
| Hiawassee River | Notteley River | Union | |
| Brasstown Creek | Hiawassee River | " | |
| Choestoe Creek | " | " | |
| Wills Creek | " | " | |

## MOBILE BASIN—IMPORTANT STREAMS

| STREAM | TRIBUTARY TO | COUNTY | REMARKS |
|---|---|---|---|
| Coosa River | Alabama River | Floyd | Formed by junction of Oostanaula and Etowah at Rome (navigable water). |
| Chattooga River | Coosa River | Chattooga | Furnishes power to Trion Factory. |
| Duck Creek | Chattooga River | Walker | |
| Silver Creek | Coosa River | Floyd | |
| Cedar Creek | " | Polk and Floyd | |
| Oostanaula River | " | Gordon and Floyd | Navigable. |
| Armuchee Creek | Oostanaula River | Chattooga and Floyd | |
| John's Creek | " | Floyd | |

| Name | Flows into | Counties | Remarks |
|---|---|---|---|
| Oothcaloga Creek | Oostanaula River | Gordon and Bartow | |
| Connasauga River | " | Whitfield and Murray | |
| Coosawattee River | " | Gilmer and Gordon | Succession of cataracts for 17 miles, from Ellijay to Carter's Mill; navigable below. |
| Sallacoa Creek | Coosawattee River | Gordon | |
| Talking Rock Creek | " | Pickens | |
| Mountain Town Creek | " | Gilmer | |
| Scared Coon Creek | " | Pickens | |
| Ellijay River | " | Gilmer | Large mountain stream. (No survey.) |
| Cartecay River | " | Gilmer | Large power at Ellijay, and others up the stream. (No survey.) Flows also through Dawson, Cherokee and Bartow Counties. |
| Etowah River | Coosa River | Lumpkin and Floyd | |
| Euharlee Creek | Etowah River | Polk and Bartow | |
| Raccoon Creek | " | Paulding | |
| Pumpkinvine Creek | " | " | |
| Allatoona Creek | " | Cobb and Bartow | |
| Little River | " | Cherokee | |
| Shoal Creek | " | " | Has one cotton factory and many undeveloped shoals. |
| Sharp Mountain Creek | " | Cherokee and Pickens | The great Marble Valley of Pickens County. See table for power. |
| Long Swamp Creek | " | Pickens | |
| Sitting Down Creek | " | Forsyth | |
| Amicalola River | " | Dawson | Amicalola Falls, 625 feet high, on head waters. See table for power. |
| Nimble Will Creek | " | Lumpkin | Source of Kin Mori mining ditch, 35 miles long. |
| Two Run Creek | " | " | |
| Shoal Creek | " | Dawson | Source of Cincinnati Consolidated mining ditch, 25 miles long, with laterals amounting to 25 miles more. |
| Mill Creek | " | Lumpkin | Source of Battle Branch mining ditch. |
| Camp Creek | " | " | |
| Jones Creek | " | " | |
| Tallapoosa River | " | Haralson | |
| Little Tallapoosa River | " | Carroll | |

## THE MOBILE BASIN—WATER-POWERS

| LOCATION OF WATER-POWER | POINT OF SECTION | STAGE | Cubic feet per second | Fall in feet | Length of Shoal | Gross Horse-power[1] | Source of Information | REMARKS |
|---|---|---|---|---|---|---|---|---|
| **BARTOW COUNTY** | | | | | | | | |
| Oothcaloga Creek | Gordon County line | Minimum | 15.0 | 6.00 | | 10.2 | Locke | |
| " | Adairsville | " | 7.0 | 6.00 | | 4.7 | " | |
| Lewis Spring | Near Adairsville | " | 8.0 | 10.00 | | 9.0 | " | |
| Cedar Spring | Martillo's Mill | " | 2.5 | 18.00 | | 5.0 | " | |
| " Creek | Gordon County line | " | 8.0 | 12.00 | | 11.0 | " | |
| Fork of Pine Log Creek | McCandless & Parrott M | " | 18.0 | 20.00 | | 41.0 | " | |
| " " | Johnson's Mill | " | 14.0 | 15.00 | | 23.8 | " | |
| Sallacoa Creek | Gordon County line | " | 20.0 | 20.00 | | 45.4 | " | |
| Stamp Creek | Pool's Furnace | " | 12.0 | 20.00 | | 27.3 | " | |
| " | At mouth | " | 24.0 | 20.00 | | 54.5 | " | |
| Boston Creek | At mouth | " | 4.0 | 20.00 | | 9.0 | " | |
| Rogers Creek | At mouth | " | 7.0 | 20.00 | | 16.0 | " | |
| Etowah River | At mouth of Allatoona Cr. | Average low water | 833.3 | 15.00 | | 1420.5 | 10th U.S. Census | |
| " | " | Average low water | 833.3 | 80.00 | | 7575.7 | " | |
| Pettis Creek | Etowah Mining Co. | Minimum | 20.0 | 5.00 | | 11.3 | Locke | |
| Nancy's Creek | At mouth | " | 6.0 | 5.00 | | 3.4 | " | |
| Two Run Creek | At mouth | " | 26.0 | 6.00 | | 17.7 | " | |
| Conaseena Creek | Kingston | " | 5.0 | 20.00 | | 11.3 | " | |
| Bamsley's Creek | Near mouth | " | 5.0 | 18.00 | | 10.2 | " | |
| Allatoona Creek | 2½ m. from mouth | " | 25.5 | 12.00 | | 49.3 | " | |
| Pumpkinvine Creek | 2 m. from mouth | " | 70.0 | 10.00 | | 79.5 | " | |
| Raccoon Creek | 1 m. from mouth | " | 39.0 | 10.00 | | 44.3 | " | |
| Euharlee Creek | 2 m. from mouth | " | 120.9 | 12.00 | | 164.8 | " | |
| **CARROLL COUNTY** | | | | | | | | |
| Little Tallapoosa River | Above mouth of Buck Cr | Low spr'g | 101.4 | 10.00 | | 115.1 | " | |
| Buck Creek | Branch of Tallapoosa | " | 16.6 | 10.00 | | 19.0 | " | |
| Indian Creek | " | " | 7.0 | 10.00 | | 7.9 | " | |
| Buffalo Creek | " | " | 6.0 | 10.00 | | 6.8 | " | |
| **CHATTOOGA COUNTY** | | | | | | | | |
| Chattooga River | Trion Factory | Ordinary | 166.6 | 16.00 | 2¾ m. | 303.0 | 10th U.S. Census | Water-power supplemented by steam for four months. |

[1] Net horse-power=80 per cent. of gross horse-power.

## MOBILE BASIN

| Stream | Location | Condition | | | | | Observer | Notes |
|---|---|---|---|---|---|---|---|---|
| **CHATTOOGA COUNTY** (Con.) | | | | | | | | |
| Armuchee Creek | Subligna | Low spr'g | 41.6 | 10.00 | .... | 47.3 | D. C. Barrow | |
| Little Turtle Creek | Near mouth | " | 5.5 | 10.00 | .... | 6.2 | " | |
| Raccoon Creek | Lot 39 | " | 4.5 | 10.00 | .... | 5.1 | " | |
| Rough Creek | At mouth | " | 8.8 | 10.00 | .... | 10.0 | " | |
| **CHEROKEE COUNTY** | | | | | | | | |
| Etowah River | Canton | Low wat'r | 733.3 | 6.25 | 6000 | 520.0 | B. M. Hall | Surveyed Aug. 27, 1890. |
| Mill Creek | " | Low spr'g | 45.0 | .... | .... | .... | D. C. Barrow | |
| Etowah River | Franklin Gold Mine | Average low water | 666.6 | 15.00 | .... | 1136.3 | 10th U. S. Census | Name now changed to Creighton Mine. |
| **DAWSON COUNTY** | | | | | | | | |
| Etowah River | Palmer's Mill | Low wat'r | 216.6 | 10.00 | .... | 246.2 | D. C. Barrow | |
| Shoal Creek | Howser's Mill | " | 33.3 | 16.00 | .... | 60.6 | " | |
| Amicalola River | Dawsonville & J. R'd | " | 150.0 | 200.00 | 17000 | 3400.0 | B. M. Hall | This is at Heard's Mill. There are other great falls below and above. |
| Amicalola Creek | Bart Crane's | Low wat'r | 10.0 | 625.00 | .... | 710.2 | " | Amicalola Falls. |
| Nimble Will Creek | Kin Mori Ditch | Ordinary | 25.0 | 300.00 | .... | 852.2 | " | At Kin Mori Mine. |
| Shoal Creek Ditch | Near Dawsonville | " | 5.0 | 200.00 | .... | 113.6 | " | Cin. Consolidated Mines. |
| **FLOYD COUNTY** | | | | | | | | |
| Etowah River | Horse Shoe Bend | .... | No measurement | No survey | .... | Said to be large power | .... | Between Rome and Kingston. |
| Armuchee Creek | Jones's Mill | Ordinary | 133.3 | 10.00 | .... | 142.3 | Locke | Little above low water. |
| Little Fork, Armuchee Cr. | Texas Valley | " | 41.0 | 15.00 | .... | 60.0 | " | Echols' Mill. |
| Big Fork | White's Bridge | " | 48.0 | 8.00 | .... | .... | " | |
| " " | Hammond's Mill | " | 48.0 | 8.00 | .... | 43.6 | " | |
| John's Creek | Near mouth | " | 15.0 | 18.00 | .... | 13.6 | " | |
| Silver Creek | " | " | 24.0 | 10.00 | .... | 49.1 | " | |
| Cedar Creek | Thoman's Mill | Minimum | 70.0 | 10.00 | .... | 79.5 | " | |
| Little Cedar Creek | Near mouth | " | 20.0 | 14.00 | .... | 32.7 | " | |
| Big Spring | Cave Springs | Low spr'g | 8.0 | .... | .... | 68.2 | " | |
| **FORSYTH COUNTY** | | | | | | | | |
| Beaver Run Creek | At mouth | Flush | 75.0 | 20.00 | .... | 170.4 | D. C. Barrow | |
| Sitting Down Creek | Hallbrook's Mill | Low spr'g | 30.0 | 7.00 | .... | 23.8 | " | |
| " " " " | Pool & Heard's | " | 30.0 | 15.00 | .... | 51.1 | " | |

## THE MOBILE BASIN—WATER-POWERS—Continued

| LOCATION OF WATER-POWER | POINT OF SECTION | STAGE | Cubic feet per second | Fall in feet | Length of Shoal | Gross Horse-power[1] | Source of Information | REMARKS |
|---|---|---|---|---|---|---|---|---|
| GORDON COUNTY | | | | | | | | |
| Oothcaloga Creek | Calhoun Mills | Low spr'g | 40.0 | 9.00 | | 40.9 | D. C. Barrow | Flat stream. |
| Connesauga Creek | At mouth | " | 291.6 | | | | " | Dam is only 9 feet, but fall is so feet in less than 2 miles. |
| Coosawattee River | Carter's Mill | " | 541.0 | 9.00 | | 562.3 | " | |
| " " | Two miles above Carter's | " | 541.0 | 50.00 | | 3073.8 | " | Heavy fall all the way. (No survey.) |
| " " | Ellijay to Carter's | " | 541.0 | | 17 m. | | " | Creek has good shoals; no survey has been made. |
| Talking Rock Creek | At mouth | " | 108.3 | | | | " | |
| Salacoa Creek | Lot 117, 7th Dist., 3d Sec. | " | 100.0 | | | | " | No fall given. |
| Snake Creek | Lot 113, 1st Dist. | " | 14.5 | | | | " | No fall given. |
| John's Creek | Lot 53, 24th Dist., 3d Sec. | " | 12.5 | | | | " | No fall given. |
| HARALSON COUNTY. | | | | | | | | |
| Tallapoosa River | Waldrop's | " | 50.0 | 10.00 | | 56.8 | " | Ten foot head assumed. |
| " " | McBride's Bridge | Flush | 583.3 | 10.00 | | 662.8 | " | Ten foot head assumed. |
| Little River | At mouth | Ordinary | 19.5 | 10.00 | | 22.1 | " | Ten foot head assumed. |
| Bench Creek | Rock House | Low wat'r | 30.5 | 30.00 | | 69.30 | " | A 30-foot dam would flood 70 acres. |
| LUMPKIN COUNTY | | | | | | | | |
| Etowah River | Five miles of Dahlonega | | 200.0 | 20.00 | | 454.5 | " | |
| " " | Simmon's Mill to Battle Branch Bridge | | | 210.00 | 10 m. | | 10th U. S. Census | |
| " " | Falls | | | 100.00 | ½ m. | | B. M. Hall | Near Cooper's Gap road. |
| Battle Branch Ditch | From Mill Creek | | 3.3 | 300.00 | | 113.6 | " | Empties into Cane Creek, to increase Hand and Barlow Mill power. |
| Etowah Ditch | From upper Etowah River | | 25.0 | 200.00 | | 568.1 | D. C. Barrow | |
| PICKENS COUNTY | | | | | | | | |
| Jones' Creek | Lot 234, 5th Dist., 1st Sec. | Low wat'r | 5.0 | 50.00 | | 28.4 | " | |
| Nimble Will Creek | 10 miles from Dahlonega | " | 50.0 | 12.00 | | 68.1 | " | |
| Big Scared Coon Creek | Fairmount Road | " | 11.0 | 10.00 | | 12.5 | " | Assumed head. |
| Talking Rock Creek | Federal Road | " | 13.3 | 10.00 | | 15.1 | " | Assumed head. |

[1] Net horse-power = 80 per cent. of gross horse-power.

MOBILE BASIN

| | | | | | | | | | |
|---|---|---|---|---|---|---|---|---|---|
| PICKENS COUNTY—(Con.) | | | | | | | | | |
| West Longswamp Creek | | " | 21.6 | 40.00 | | | 98.4 | B. M. Hall | Perseverance Marble Quarries. Surveyed Jan., 1890. |
| East Longswamp Creek | | " | 6.6 | 50.00 | 2,600ft | | 94.7 | " | Pelton wheel, one mile ditch, and 1,500 foot pipe. |
| Rocky Creek | | " | 3.6 | 210 00 | | | 87.5 | " | Fall about 30 feet in one mile. |
| Long Swamp Creek | Georgia Marble Co. | " | 46.6 | | | | . . | " | Surveyed Nov., 1890. |
| " | Blue Ridge Marble Co | " | 50.7 | 16.00 | 3,200ft | | 92.1 | " | |
| POLK COUNTY | | | | | | | | | |
| Euharlee Creek | Rockmart | Low wat'r | 25.0 | 10.00 | | | 28.4 | D. C. Barrow | |
| " | 2 miles north of Rockmart | Low spr'g | 19.0 | 10.00 | | | 21.6 | " | |
| " | Hightower's Mill | " | 5.4 | 90.00 | | | 55.2 | " | |
| Big Spring | 2 miles from Van Wert | " | 5.0 | | | | . . | " | |
| Little Cedar Creek | Young's Mill | " | 19.3 | 10.00 | | | 20.7 | " | |
| Big Spring | Cedartown | " | 9.6 | | | | . . | " | |
| Gut Creek | At mouth | " | 26.6 | 10.00 | | | 30.3 | " | Assumed head of 10 feet. |
| PAULDING COUNTY | | | | | | | | | |
| Little Pumpkinvine Creek | 16 miles from Marietta | " | 10.0 | 20.00 | | | 22.7 | Locke | |
| Raccoon Creek | Chappel's Store | " | 22.0 | 12.00 | | | 30.0 | " | |
| WALKER COUNTY | | | | | | | | | |
| Fork of Dry Creek | One-half mile from mouth | " | 6.5 | 10.00 | | | 7.3 | D. C. Barrow | |
| WHITFIELD COUNTY | | | | | | | | | |
| Swamp Creek | Lot 113 | " | 33.3 | 10.00 | | | 37.8 | " | Assumed head. |
| Carpenter Creek | One-half mile So. of Tilton | " | 11.0 | 10.00 | | | 12.5 | " | Assumed head. |
| Mill Creek | Lot 148, 13th Dist., 3d Sec. | " | 16.0 | 10.00 | | | 18.1 | " | Assumed head. |
| Etowah River | For 17 miles above W. & A. bridge | Low wat'r | 833.3 | 102 00 | 17 m. | | 9,650.0 | . . | From mouth of Little River in Cherokee Co. to W. & A. R. R. bridge in Bartow Co. |
| " | Cartersville to Rome | " | 1000.0 | 154 00 | 45 m. | | 17,500.0 | | |

The foregoing gives a very meagre idea of the water-powers of this basin. The surveys made by Messrs. Barrow and Locke, Assistant State Geologists, in 1874-'75, were confined mainly to that part of the basin, in which the streams have very few shoals of importance. The great shoals on the Coosawattee, the Cartecay and the Amicalola rivers, and the head streams of the Etowah River, have as yet received very little attention.

## MOBILE BASIN — UTILIZED POWER

| STREAM | COUNTY | KIND OF MILL | No. of Mills | Total fall used | Total H. P. used | REMARKS |
|---|---|---|---|---|---|---|
| Tallapoosa River | Haralson | Flour and grist | 3 | 16 | 67 | |
| " | " | Saw | 1 | 7 | 12 | |
| " | Paulding | Flour and grist | 1 | 10 | 10 | |
| Tributaries of Tallapoosa R. | Haralson | " | 7 | 71 | 92 | |
| " | " | Saw | 1 | 6 | 5 | |
| " | Carroll | Cotton gin | 1 | 6 | | |
| " | " | Flour and grist | 10 | 142 | 151 | |
| " | " | Saw | 3 | 32 | 36 | |
| " | " | Tannery | 1 | 24 | 6 | |
| " | " | Woolen | 2 | 20 | 9 | |
| Tributaries of Coosa River | Floyd | Flour and grist | 14 | 183 | 204 | |
| " | " | Saw | 3 | 37 | 43 | |
| " | " | Woolen | 2 | 15 | 17 | |
| " | Polk | Cotton gins | 2 | 23 | 20 | |
| " | " | Machine shop etc. | | 4 | 70 | |
| " | " | Flour and grist | 6 | 125 | 138 | |
| " | " | { Flour and grist, saw and tannery. } | | | | Cedartown. |
| Etowah River | Dawson | Stamp mill | 1 | 18 | 58 | |
| " | " | Flour and grist | 4 | 83 | 50 | |
| " | " | Saw | 2 | 42 | 27 | |
| Tributaries of Etowah River | Polk | Flour and grist | 2 | 30 | 40 | |
| " | Floyd | " | 2 | 16 | 47 | |
| " | Bartow | " | 14 | 156 | 308 | |
| " | Paulding | " | 9 | 107 | 79 | |
| " | " | Saw | 2 | 24 | 34 | |
| " | " | Woolen | 1 | 12 | 4 | |
| " | Cobb | Flour and grist | 2 | 26 | 26 | |
| " | Cherokee | " | 12 | 195 | 187 | |
| " | " | Cotton gins | 2 | 25 | 56 | |
| " | Pickens | Saw | 5 | 78 | 64 | |
| " | " | " | 5 | 54 | 50 | |

# MOBILE BASIN

| Basin | County | Industry | | | |
|---|---|---|---|---|---|
| Cooswattee R. and Trib'r's | " | Furniture | 2 | 15 | 20 |
| " | " | Flour and grist | 13 | 179 | 129 |
| " | " | Marble mill | 1 | 210 | 60 |
| " | Milton | Flour and grist | 2 | 28 | 16 |
| " | " | Wheelwright | 1 | 12 | 6 |
| " | " | Saw | 4 | 68 | 74 |
| " | Dawson | Flour and grist | 2 | 38 | 40 |
| " | " | Woolen | 1 | | 8 |
| " | Bartow | Flour and grist | 5 | 57 | 74 |
| " | Gilmer | " | 3 | 61 | 48 |
| " | Gordon | " | 5 | 41 | 160 |
| " | " | Cotton gin | 1 | | |
| " | " | Saw | 1 | | |
| " | " | Tannery | 1 | 18 | 432 |
| " | Pickens | Cotton factory | 10 | 141 | 116 |
| " | " | Flour and grist | 1 | 12 | 10 |
| " | " | Saw | 1 | 18 | 20 |
| " | " | Woolen | 8 | 93 | 105 |
| Conasauga R. and Trib'r's Murray | " | Flour and grist | 2 | 28 | 30 |
| " | " | Saw | 2 | 20 | 22 |
| " | Whitfield | Flour and grist | 12 | 161 | 151 |
| " | " | " | 1 | 7 | 6 |
| " | " | Boots and shoes | 4 | 56 | 52 |
| " | Bartow | Flour and grist | 2 | 17 | 13 |
| " | " | Woolen | 1 | 10 | 7 |
| " | Chattooga | Cotton gin | 3 | 43 | 50 |
| " | " | Flour and grist | 6 | 74 | 122 |
| " | " | " | 3 | 24 | 141 |
| " | Floyd | Saw | 3 | 43 | 24 |
| " | Gordon | Flour and grist | 1 | 16 | 300 |
| Chattooga R. and Trib'r's Chattooga | " | Cotton factory | 4 | 24 | 40 |
| " | " | " gin | 7 | 92 | 145 |
| " | " | Flour and grist | 5 | 63 | 102 |
| " | " | Saw | 1 | 13 | 8 |
| " | " | Woolen | | | |

## APALACHICOLA BASIN — IMPORTANT STREAMS

| NAME OF STREAM | TRIBUTARY TO | COUNTY | REMARKS |
|---|---|---|---|
| Chattahoochee River | Apalachicola River | | |
| Standing Boy Creek | Chattahoochee | Muscogee | Large shoal on creek, 2 m. from mouth. |
| Mulberry Creek | " | Harris | Large cr.; falls 60 ft. in quarter of mile. |
| Mountain Creek | " | " | |
| Old House Creek | " | " | 60 cu. ft. per sec.; 20 ft. fall on shoal at River Road. |
| Flat Shoals | " | Harris and Troup | Troup Factory, 80 cu. ft. per sec.; 18 ft. fall, low water. (Locke) |
| Muddy Creek | " | Troup | 5½ m. from LaGrange; 7 cu. ft. per sec.; 10 ft. fall, low water. (Locke) |
| Yellow Jacket Creek | " | " | 8½ m. from LaGrange; 87 cu. ft. per sec.; 10 ft. fall, low water. (Locke) |
| Beach Creek | Yellow Jacket Creek | " | 5 m. from LaGrange; 35 cu. ft. per sec.; 15 ft. fall, low water. (Locke) |
| Panther Creek | Chattahoochee River | " | 3 m. from LaGrange; 25 cu. ft. per sec.; 10 ft. fall, low water. (Locke) |
| Flat Creek | " | " | Gorham's Mill; 20 cu. ft. per sec.; 12 ft. fall, low water. (Locke) |
| New River | " | Heard and Coweta | ¼ m. of mouth; 133.3 cu. ft. per sec.; 10 ft. fall, low spring. (Locke) |
| Whittaker Creek | " | Heard | Whittaker's Mill; 91 cu. ft. per sec.; 30 ft. fall. (C. C. Anderson) |
| Hillabahatchee Creek | " | " | |
| Centralhatchee Creek | " | " | 57.9 cu. ft. per sec.; 8 ft. fall, saw mill. (C. C. Anderson) |
| Wahoo Creek | " | Coweta | At Sergeant's; 41.4 cu. ft. per sec. at mean low water. Fall, 33 ft. in 1,600. (C. C. Anderson) Cotton factory and grist mill. |
| Cedar Creek | " | " | |
| Snake Creek | " | Carroll | 2.6 cu. ft. per sec, 14 ft. fall=$\frac{3}{16}$ H. P. per foot of fall. (C. C. Anderson) |
| Dog River | " | " | Above Watkin's mill; 25 cu. ft. per sec., low spring. (Locke) |

# APALACHICOLA BASIN

| | | | |
|---|---|---|---|
| Bear Creek | " | Douglas | 52.5 cu. ft. per sec. (C. C. Anderson) |
| Camp Creek | " | Campbell | |
| Sweet Water Creek | " | Paulding, Cobb and Douglas | Austell Shoals, near mouth, has 80 feet of fall and 166.9 cu. ft. per sec. Hayes bridge, 80 cu. ft. per sec., low water. (Locke) |
| Powder Springs Creek | Sweet Water Creek | Cobb | Powder Springs; 34 cu. ft. per sec., low water. (Locke) |
| Nose's Creek | " | " | |
| Soap Creek | Chattahoochee River | Fulton | Paper mill; 62 cu. ft. per sec.; 67 ft. head, low spring. (Locke) |
| Utoy Creek | " | " | |
| Nickajack Creek | " | Cobb | 29 ft. fall at Ruff's Mill, and 21 ft. at Concord Factory. |
| Peachtree Creek | " | Fulton and DeKalb | Houston's Mill; 23.3 cu. ft. per sec; 22 ft. fall, low water. (Locke) Buckhead Road, 97 cu. ft. per sec., flush. (Locke) |
| Nancy's Creek | Peachtree Creek | " | Lot 96, 17th Dist.; 45 cu. ft. per sec., low spring. (Locke) |
| Rottenwood Creek | Chattahoochee River | Cobb | 12 cu. ft. per sec. = 1.27 gross H.P. per ft. of fall; measured July 28, 1892, by B. M. Hall. |
| Long Island Creek | " | Fulton | Lot 164, 17th Dist., 6.5 cu. ft. per sec. (Locke) |
| Willis Creek | " | Cobb | Wright's Mill; 16.6 cu. ft. per sec.; 23 ft. fall, ordinary stage; gross H.P. = 43. |
| Vickery's Creek | " | Forsyth, Milton and Cobb | 3 factories at Roswell; total fall, 103 ft.; volume about 50 cu. ft. per sec. (C. C. Anderson) |
| Suwanee Creek | " | Gwinnett | Lawrenceville and Buford road; 11.6 cu. ft. per sec. (Locke) |
| Ivy Creek | Suwanee Creek | " | Hamilton's Mill; 2 cu. ft. per sec., 18 ft. fall, low water. (Locke) |
| Chestatee River | Chattahoochee River | Lumpkin, Dawson, Forsyth and Hall | Important gold mining stream, with many fine undeveloped powers, not surveyed. |
| Etowah Ditch, entering Cane Creek | Chestatee River | Lumpkin | Ditch, 7 miles long, diverts Etowah waters across ridge into Cane Creek; 25 cu. ft. per sec., with a head of 200 ft. = 568 gross H.P.; not utilized. |
| Cane Creek | " | " | At Cane Cr'k falls, 16.6 cu. ft. per sec.; 60 ft. fall. At Barlow gold-mill, 40 cu. ft. per sec. |

## APALACHICOLA BASIN — IMPORTANT STREAMS — *Continued*

| NAME OF STREAM | TRIBUTARY TO | COUNTY | REMARKS |
|---|---|---|---|
| Clay Creek | Cane Creek | Lumpkin | Has a good shoal. |
| Yahoola Creek | Chestatee River | " | Source of Hand Mining Ditch, 35 miles long; furnishes water to many mines for hydraulic mining. The ditch carries from 16 to 25 cu. ft. per sec.; and is 300 ft. above streams near Dahlonega. |
| Cavender's Creek | " | " | Drains an important gold-mining region of Lumpkin county. |
| Yellow Creek | " | Hall | 7.2 cu. ft. per sec.; 20 ft. shoal near mouth. (Barrow) |
| Tessantee River | " | White | 95 cu. ft. per sec.; big shoal near mouth. |
| Shoal Creek | Tessantee River | " | Has Asbury's Mill and other good shoals. |
| Town Creek | " | " | Source of Loud Ditch, 25 miles long, used for hydraulic mining. |
| Jennie's Creek | Town Creek | " | |
| Tate's Creek | Chestatee River | Lumpkin | To furnish water for proposed Cavender's Creek Ditch. |
| Mill Creek | " | " | To furnish water for proposed Cavender's Creek Ditch. |
| Dick's Creek | " | " | |
| Turner's Creek | " | White | Large creek; falls over 100 feet to the mile. |
| Little R. from Wahoo Cr. | Chattahoochee River | Hall | Castleberry's Mill, 4 miles from Gainesville; 151.5 cu. ft. per sec.; 71 ft. fall; gross H. P., 122; 25 H. P. used. (C. C. Anderson) |
| Glade Creek | " | " | Furnishes water and drainage to "The Glades" Gold Mine. |
| Flat Creek | " | " | 13.6 cu. ft. per sec.; 50 ft. fall; shoal above "The Glades" Mine. |
| Mud Creek | " | Habersham | Big Mud Creek, 33.3 cu. ft. per sec.; Little Mud Creek, 20 cu. ft. per sec. |
| Soquee River | " | " | See Power Table. |
| Hazel Creek | Soquee River | " | Lake and water-power at Demorest. |
| Deep Creek | " | " | 38.3 cu. ft. per sec. at mouth. (Barrow and Locke) |

# APALACHICOLA BASIN

| Stream | Tributary to | County | Remarks |
|---|---|---|---|
| Shoal Creek | " | " | 16.6 cu. ft. per sec. at mouth. (B. M. Hall, estimated) |
| Mossy Creek | Chattahoochee River | White | |
| Duke's Creek, North Fork | " | " | Duke's Creek Falls, 12.8 cu. ft. per sec.; 300 ft. fall. (Barrow) |
| " | " | " | Minnehaha Falls, 3.6 cu. ft. per sec.; 300 ft. fall. (Barrow) |
| Smith's Creek | " | " | Annie Ruby Falls, 7.1 cu. ft. per sec.; 300 ft. fall. (Barrow) |
| Flint River | Apalachicola River | Webster, Sumter, Terrell | Large Creek with fine undeveloped power, enough for running 100,000 spindles. (U.S. Government Report) |
| Kinahatoochee Creek | Flint River | | |
| Buck's Creek | " | Macon | |
| Whitewater Creek | " | Macon and Taylor | |
| Cedar Creek | Whitewater Creek | Taylor | |
| Parchelagee Creek | Flint River | " | |
| Spring Creek | " | Crawford | |
| Little Potato Creek | " | Upson | |
| Big Potato Creek | " | Upson and Pike | |
| Wasp Creek | Big Potato Creek | Pike | |
| Grape Creek | " | " | |
| Laser Creek | Flint River | Talbot | |
| Pigeon Creek | " | Meriwether and Talbot | |
| Cane Creek | " | Meriwether | |
| Red Oak Creek | " | " | |
| Elkin's Creek | " | Pike | |
| Line Creek | " | Coweta and Fayette | |
| Whitewater Creek | Line Creek | Fayette | |

## APALACHICOLA BASIN — WATER-POWERS

| Utilized Net H.P. | LOCATION OF WATER-POWER | POINT OF SECTION | Stage of Water | Cubic ft. per Second | Fall in feet | Length of shoal | Gross H.P.[1] | Source of Information | REMARKS |
|---|---|---|---|---|---|---|---|---|---|
| | **SOQUEE RIVER** | | | | | | | | |
| 60 | Habersham County | Clarkesville Woolen Mill | 0.0 | 266.6 | 26.0 | 1,000' | 738.6 | C.C. Anderson | Only 18 ft. used. |
| 100 | " | Porter Mills, Shoal No. 1 | " | 266.6 | 14.4 | 1,00' | 436.3 | " | See fluctuation tables; 0.0 = min. observed waters. |
| 150 | " | Porter Mills, Shoal No. 2 | " | 291.6 | 45.2 | 1,400' | 1,369.0 | " | |
| None | " | Porter Mills, Shoal No. 3 | " | " | 15.0 | 1,200' | 497.0 | " | |
| | **CHATTAHOOCHEE RIVER** | | | | | | | | |
| Corn Mill | White County | Nicholls' Mill | Min. L.W. | 72.0 | 10.0 | . . . | 81.8 | Barrow & Locke | |
| None | White & Habersham Cos. | Duncan Shoal | 0.0 | 683.3 | 7.6 | 400' | 589.2 | C.C. Anderson | Includes Soquee River at mouth. |
| " | " | Carpenter Shoal | " | 683.3 | 3.2 | 400' | 248.4 | " | Below mouth of Soquee. |
| " | " | Johnny's Ford Shoal | " | 683.3 | 5.4 | 1,200' | 419.3 | " | |
| " | " | Gearing Shoal | " | 683.3 | 1.3 | 300' | 101.0 | " | |
| " | " | Fishtrap Shoal | " | 683.3 | 1.8 | 300' | 138.8 | " | |
| " | " | Bull Shoal | " | 683.3 | 7.0 | 1,800' | 543.5 | " | Foot, 3 miles below mouth of Soquee. Can be developed as one power. |
| " | " | Last Six Shoals, total | " | 683.3 | 38.0 | 13,200' | 2,950.7 | " | |
| " | " | Rock House Shoal | " | 750.0 | 3.7 | 900' | 315.3 | " | |
| " | " | Mountain Island Shoal | " | 766.6 | 7.3 | 1,800' | 635.8 | " | |
| " | Hall County | Lula Bridge | " | 783.3 | 2.0 | 1,200' | 178.0 | " | |
| " | " | Reynolds | " | 800.0 | 6.0 | 1,200' | 545.4 | " | |
| " | " | Seven Islands | " | 816.6 | 4.0 | . . . | 371.2 | " | |
| " | " | Savage Shoal, No. 1 | " | 833.3 | 1.0 | 1,200' | 94.7 | " | |
| " | " | Savage Shoal, No. 2 | " | 833.3 | 2.5 | 1,200' | 236.7 | " | |
| " | " | Peg's Shoal | " | 833.3 | 6.3 | 2,530' | 596.0 | " | |
| " | " | Stringer's Ford | " | 833.3 | 10.0 | 1,200' | 947.0 | " | |
| " | " | Wilson Shoal | " | 933.3 | 6.5 | 2,500' | 689.4 | " | |
| " | " | Thompson's Bridge | " | 933.3 | . | . . . | . . . | " | |

[1] Net H. P. = 80 per cent. of gross H. P.

WATER-POWERS OF GEORGIA

PLATE III

HURRICANE FALLS, TALLULAH FALLS, GEORGIA.

APALACHICOLA BASIN

| County | Shoal | | | | | Surveyor | Remarks |
|---|---|---|---|---|---|---|---|
| " | Shallow Ford | 933.3 | 6.70 | 5,500' | 710.6 | U.S. Sur. | Vol. estimated from Sur. of C. C. Anderson. |
| " | Johnson's Shoal | 933.3 | 3.20 | 3,600' | 339.4 | " | " |
| " | Mooney's Shoal | 933.3 | 3.20 | 5,600' | 339.4 | " | Below Mouth of Chestatee. |
| " | Overby's Shoal | 1,450.0 | 6.90 | 800' | 1,137.0 | " | Vol. estimated from Sur. of C. C. Anderson. |
| Mill and Gin, 80 | Brown's Bridge | 1,450.0 | 17.00 | 8,500' | 2,801.0 | " | |
| " | Firkle Shoal | 1,450.0 | 3.90 | 4,000' | 642.3 | " | |
| " | Garner's Shoal | 1,666.6 | | 1,182' | | " | |
| Gwinnett County | Bridge Shoal | 2,000.0 | 16.90 | | 3,841.0 | " | |
| " | Jones's Shoal | 2,083.3 | 3.10 | 1,200' | 733.9 | " | |
| Milton County | Island Ford Shoal | 2,133.3 | 9.00 | 5,000' | 2,181.0 | " | |
| None | Roswell Shoal | 2,190.5 | 18.00 | about 2 mi. | 4,480.0 | Anderson | From Bridge to head of Bull Sluice. |
| Cobb and Fulton Co's [1] | Bull Sluice Shoal | 2,200.0 | 25.30 | 1 mile | 6,325.0 | " | On Pink Power Property. |
| " | " continued | 2,200.0 | 6.40 | 3,300' | 1,660.0 | " | On Strapp & Power " |
| " | Cochran Shoal | 2,333.3 | 6.50 | 2,700' | 1,723.0 | " | Above Power's Ferry. |
| " | Devil's Race Course | 2,333.3 | 10.50 | 2,500' | 2,784.0 | " | Below "The Narrows." |
| " | Upper Thornton Shoal | 2,333.3 | 4.00 | 1,100' | 1,219.0 | " | Head of Island to Little Nancy's Creek. |
| " | Long Island Shoal | 2,358.3 | 10.00 | 5,900' | 2,679.0 | " | Includes the four shoals above. |
| " | Top of Cochran Shoal to foot of I. I. Shoal | 2,358.3 | 32.80 | 18,100' | 8,790.0 | Hall | |
| " | Howell's Shoal | 2,366.6 | 10.70 | 4,000' | 2,877.0 | Anderson | |
| " [2] | W. & A. R. R. Bridge | 2,500.0 | | | | | |
| Campbell County | Redman's Shoal | 2,500.0 | 3.00 | 1,000' | 848.0 | Anderson | |
| " | Pumpkintown Shoal | 2,666.6 | 3.00 | 800' | 909.0 | " | |
| " | Mederis Shoal | 2,666.6 | 8.40 | 2,000' | 2,545.4 | " | |
| Coweta County | Island Shoal | 2,750.0 | 12.50 | 5,280' | 3,996.0 | " | |
| " | Fridell Shoal | 2,750.0 | 9.00 | 1,400' | 2,812.5 | " | |
| " | McIntosh Shoal | 2,833.3 | 11.62 | 19,000' | 3,741.0 | " | Fall by B. M. Hall. |
| Heard County | Hilly Mill | 2,833.3 | 7.00 | 2,000' | 2,632.5 | " | |
| " | Bush Head Shoal | 2,916.6 | 5.00 | 1,000' | 1,657.0 | " | |
| 50 H. P. | Hendrick's Shoal | 2,916.6 | 16.50 | 4,000' | 5,468.7 | " | Grist-mill. |

[1] These three shoals form one continuous shoal four miles long with a fall of fifty feet.

[2] Known as the Vining Shoal, being near Vining Station on W. & A. R. R.

## APALACHICOLA BASIN — WATER-POWERS — Continued

| Utilized Net H.P. | LOCATION OF WATER-POWER | POINT OF SECTION | Stage of Water | Cubic Feet per Second | Fall in Feet | Length of Shoal | Gross H. P.[1] | Source of Information | REMARKS |
|---|---|---|---|---|---|---|---|---|---|
| None | Heard County | Jackson Shoal | 0.0 | 3,066.6 | 6.7 | 3,000' | 2,296.7 | Anderson | |
| " | " | Seven small Shoals | " | 3,333.3 | 13.0 | | 4,924.0 | " | |
| " | Troup County | Swanson Shoal | " | 3,500.0 | 7.0 | 1,500' | 2,784.0 | " | |
| " | " | Small Shoals | " | 3,750.0 | 3.5 | | 1,491.5 | " | |
| " | " | McGees' Bridge | " | 4,000.0 | 8.3 | 3,000' | 3,772.7 | " | |
| " | " | Buzzard and Reed Island | " | 4,166.6 | 8.3 | 3,000' | 3,930.0 | " | Three shoals. |
| " | " | Bentley's Mill | " | 4,166.6 | 4.0 | | 1,894.0 | " | |
| " | " | Ferrell or Huguley's | " | 4,666.6 | 9.0 | | 4,772.7 | " | |
| " | " | Pott's Shoal | " | 4,933.3 | 5.0 | 3,600' | 2,803.0 | " | 3 or 4 miles above W. P. |
| " | " | West Point | " | 4,933.3 | | | | " | |
| 300H.P. | Harris County | Jack Todd's Shoal | " | 4,933.3 | 51.0 | 39,600' | 28,591.0 | U. S. Sur. | Two cotton-mills, four miles below W. P. Vol. from C. C. Anderson |
| " | " | 3 m. below Houston's Ferry | " | 4,933.3 | 4.0 | 1,100' | 2,242.0 | " | |
| None | " | Hargett's Island Shoal | " | 5,000.0 | 60.0 | 13,000' | 34,091.0 | " | |
| " | " | Shoal | " | 5,000.0 | 15.0 | 4,000' | 8,522.7 | " | |
| " | " | Tate Shoals | " | 5,000.0 | 26.0 | 8,700' | 14,772.0 | " | |
| " | " | Mulberry Shoals | " | 5,000.0 | 22.0 | 6,300' | 12,500.0 | " | |
| " | Muscogee County | Near mouth of Standing Boy Creek | " | 5,166.6 | 30.0 | 10,560' | 17,613.0 | " | |
| " | " | Chattahoochee Falls Prop. | " | 5,216.6 | 10.0 | 3,800' | 5,928.0 | " | |
| " | At Columbus | Lover's Leap | " | 5,216.6 | 42.0 | 6,900' | 24,715.0 | " | |
| " | " | City Mills | " | 5,216.6 | 37.0 | 2,600' | 21,933.0 | " | |
| " | " | Eagle and Phœnix Mills | " | 5,216.6 | 10.0 | Dam | 5,928.0 | " | |
| " | Hall, Bartow, Muscogee and Intervening Counties | | " | 5,216.6 | 25.0 | " | 14,820.0 | " | |
| | Continuous level from Thompson's Bridge to W. & A. Ry. Bridge | | " | | 227.0 | 73 miles | | " | 3 m. N. of Gainesville to 6 m. W. of Atlanta. |
| | From W. & A. Ry. Bridge to West Point | | " | | 162.0 | 108 mls. | | " | 6 m. W. of Atlanta to West Point. |
| | From West Point to Columbus | | " | | 362.0 | 34 mls. | | " | W. Point to Columbus. |

[1] Net horse-power = 80 per cent. of gross horse-power.

APALACHICOLA BASIN 35

| | | | | | | | | | |
|---|---|---|---|---|---|---|---|---|---|
| **SWEETWATER CREEK** | | | | | | | | | |
| Douglas County | Austell Shoals | | LowWat'r | 166.6 | 80.0 | 3,900' | 1,515.0 | B. M. Hall | Near Austell, Ga. Easily developed. |
| **CHESTATEE RIVER** | | | | | | | | | |
| Lumpkin County | Garnet Mine | | " | Unk'n | 15.0 | 1,200' | Unk'n | " | Dam, race, stamp-mill and pumps. |
| " | Chestatee Pyrites Co. | | " | " | 20.0 | Unk'n | " | " | |
| " | Penitentiary Shoal | | " | " | L'rge | " | " | " | Power developed. |
| " | Chestatee Mining Co. | | " | " | Unk'n | " | " | " | |
| " | Calhoun Mine | | " | 290.0 | 12.0 | Dam | 395.0 | Barrow | Dam, stamp-mill and pump. |
| " | Leather's Ford | | " | | 12.0 | Unk'n | | | |
| **FLINT RIVER** | | | | | | | | | |
| Meriwether and Pike Cos. | Sullivan's Mill | 30 H.P. | 0.0 | 250.0 | 7.3 | 200' | 207.0 | Anderson | Grist-mill. |
| " | | 40 " | Min.L.W. | 258.3 | 32.0 | 3,000' | 934.0 | B. M. Hall | A four foot storage-dam will develop 2,630 gross 10 hour H. P., 6 days per week, at lowest water. |
| " | Flat Shoals | | | | | | | | |
| Upson County | Dripping Rock | None | Normal Flush | 856.6 | 32.0 | 3,000' | 3,114.0 | Anderson | Water too high for measurement. |
| " | Yellow Jacket Shoals | " | Normal Flush | 1,674.1 1,216.2 | 14.0 36.6 | 2,900' 3,400' | | | |
| " | Snipe's Shoals | " | | 2,607.6 | 7.0 | 1,800' | | Anderson | |
| **BIG POTATO CREEK** | | | | | | | | | |
| Upson County | Rogers' Shoals | None | LowWat'r | 103.3 | 81.0 | 3,500' | 951.0 | " | |
| " | Nelson's Shoals | 30 H.P. | 0.0 | 110.0 | 115.0 | 2,700' | 1,437.0 | " | 1st drop is 60 ft, in a distance of 500 ft, making 750 gross H. P. |
| " | Daniel's Mill | 30 " | " | 110.0 | 13.0 | 150' | 162.0 | " | |
| **CHATTAHOOCHEE CO.** | | | | | | | | | |
| Oswitchee Creek | Romney's Mill | | Low Sp'g | 21.0 | 18.0 | | 42.0 | Locke | |
| Woolfolk's Branch | Woolfolk's | | | 1.0 | 65.0 | | 7.0 | | |
| **CLAY COUNTY** | | | | | | | | | |
| Chemochechobee Creek | Weaver's Mill | | " | 60.0 | 30.0 | | 204.0 | Barrow | |
| Pataula Creek | Rapids | | " | 240.0 | 22.0 | | 600.0 | " | |

## APALACHICOLA BASIN — WATER-POWERS — *Continued*

| Utilized Net H.P. | Location of Water-Power | Point of Section | Stage of Water | Cubic Feet per Second | Fall in Feet | Length of Shoal | Gross H.P.[1] | Source of Information | Remarks |
|---|---|---|---|---|---|---|---|---|---|
| | **DECATUR COUNTY** | | | | | | | | |
| .. | Limesink Creek .... | Limesink ........ | Low Spr'g | 2.0 | 105.0 | .. | 23.0 | Locke | Creek disappears. |
| .. | Barnett's Creek .... | Lot 367 ........ | " | 23.0 | 10.0 | .. | 26.0 | " | Flow affected by mills above. |
| .. | Attapulgas Creek ... | Thomasville Road .. | " | 18.0 | .. | .. | .. | " | |
| .. | Sanburn's Creek .... | Attapulgas Road .. | " | 8.0 | .. | .. | .. | " | |
| | **EARLY COUNTY** | | | | | | | | |
| .. | Harrod's Creek .... | Early Factory .... | " | 20.0 | 35.0 | .. | 80.0 | " | |
| .. | Colomochee Creek .. | Early Road ...... | " | 70.0 | 12.0 | .. | 95.0 | " | |
| | **QUITMAN COUNTY** | | | | | | | | |
| .. | Heclamee Creek .... | Near Mouth ...... | Low Wat'r | 6.0 | 10.0 | .. | 7.0 | " | |
| .. | Tobehannee Creek .. | " Georgetown .. | " | 10.0 | 10.0 | .. | 11.0 | " | |
| | **RANDOLPH COUNTY** | | | | | | | | |
| .. | Roaring Branch .... | Five miles from Fort Gaines | " | 4.0 | 30.0 | .. | 14.0 | " | |
| .. | Wakefortsee Creek .. | Near Chemochechobee .. | " | 5.0 | 10.0 | .. | 5.0 | " | |
| | **STEWART COUNTY** | | | | | | | | |
| .. | Wimberly's Branch .. | Gaines & Freeman's Mill .. | " | 8.8 | 12.0 | .. | 12.0 | " | |
| .. | Hodchodkee Creek .. | Scott's Mill ...... | " | 12.0 | 10.0 | .. | 14.0 | " | |

Many important water-powers are omitted in the Apalachicola Basin for want of data. The foregoing is the best that can be done, until more surveys are made. Investigation is especially needed on the Flint River and its upper tributaries.

[1] Net H. P. = 80 per cent. of Gross H. P

## APALACHICOLA BASIN — UTILIZED POWER

| STREAM | COUNTY | KIND OF MILL | No. of Mills | Total Fall Used, in Feet | Total Net H.P. Used | REMARKS |
|---|---|---|---|---|---|---|
| Chattahoochee River | Muscogee | Cotton Factories | 3 | 43 | 2,000 | |
| " | " | Flour and Grist | 1 | 8 | 100 | |
| " | Harris | " | 1 | 8 | 50 | |
| " | " | Cotton Factory | 1 | 8 | 160 | |
| " | Troup | " | 1 | 9 | 130 | |
| " | Hall | Building Material | 1 | 9 | 30 | |
| " | " | Flour and Grist | 1 | 9 | 60 | |
| " | Cobb | " | 1 | 11 | 10 | |
| Trib'y's of Chat'h'chee River | Early | Sawmill | 6 | 56 | 72 | |
| " | " | " | 1 | . | 25 | |
| " | Clay | " | 3 | 29 | 60 | |
| " | " | Cotton Gin | 1 | . | 6 | |
| " | " | Flour and Grist | 6 | 58 | 77 | |
| " | Quitman | " | 4 | 49 | 96 | |
| " | " | Sawmill | 2 | 24 | 63 | |
| " | Randolph | Flour and Grist | 1 | 9 | 8 | |
| " | Stewart | " | 8 | 83 | 192 | |
| " | " | Sawmill | 2 | 20 | 22 | |
| " | Chattahoochee | " | 1 | 10 | 15 | |
| " | " | Flour and Grist | 6 | 57 | 75 | |
| " | Muscogee | " | 4 | 73 | 213 | |
| " | Marion | " | 1 | 6 | 12 | |
| " | " | Cotton Gin | 1 | 8 | 21 | |
| " | " | Sawmill | 1 | 8 | 30 | |
| " | Harris | " | 1 | 12 | 10 | |
| " | " | Flour and Grist | 13 | 235 | 398 | |
| " | Talbot | " | 2 | 36 | 47 | |
| " | " | Sawmill | 2 | 36 | 43 | |
| " | Troup | " | 4 | 57 | 65 | |
| " | " | Tannery | 1 | 22 | 8 | |
| " | " | Flour and Grist | 22 | 223 | 506 | |
| " | " | Cotton | 1 | 20 | 60 | |

## APALACHICOLA BASIN — UTILIZED POWER — *Continued*

| STREAM | COUNTY | KIND OF MILL | No. of Mills | Total Fall Used, in Feet | Total Net H.P. Used | REMARKS |
|---|---|---|---|---|---|---|
| Trib'r's of Chatt'ch'ee River | Meriwether | Flour and Grist | 1 | 30 | 11 | |
| " | Heard | " | 8 | 91 | 101 | |
| " | " | Sawmill | 3 | 124 | 125 | |
| " | Carroll | Cotton | 1 | 30 | 120 | |
| " | " | Flour and Grist | 12 | 277 | 160 | |
| " | " | Sawmill | 3 | 58 | 26 | |
| " | Coweta | Cotton | 1 | | 60 | |
| " | " | Flour and Grist | 14 | 275 | 226 | |
| " | Campbell | " | 7 | 124 | 130 | |
| " | Douglas | Cotton Gin | 1 | 11 | 20 | |
| " | " | Flour and Grist | 13 | 202 | 119 | |
| " | " | Sawmill | 6 | 136 | 82 | |
| " | " | Tannery | 1 | 60 | 10 | |
| " | " | Cotton | 1 | | 60 | |
| " | Paulding | Woolen-mill | 1 | 14 | 9 | |
| " | " | Flour and Grist | 2 | 13 | 60 | |
| " | " | Sawmill | 1 | 20 | 8 | |
| " | Cobb | Cotton | 3 | 67 | 375 | |
| " | " | Woolen-mill | 2 | 40 | 85 | |
| " | " | Cotton Gins | 9 | 135 | 111 | |
| " | " | Flour and Grist | 23 | 368 | 454 | |
| " | " | Paper-mill | 1 | 22 | 75 | |
| " | " | Sawmill | 5 | 45 | 69 | |
| " | Fulton | " | 3 | 30 | 31 | |
| " | " | Cotton Gins | 3 | 20 | 22 | |
| " | " | Flour and Grist | 8 | 157 | 106 | |
| " | DeKalb | " | 7 | 120 | 119 | |
| " | " | Furniture | 2 | 47 | 25 | |
| " | " | Tannery | 1 | 15 | 10 | |
| " | " | Sawmill | 2 | 24 | 40 | |
| " | Gwinnett | " | 4 | 47 | 44 | |
| " | " | Flour and Grist | 9 | 116 | 98 | |

## APALACHICOLA BASIN

| | | | | | | |
|---|---|---|---|---|---|---|
| Forsyth | " | Sawmill | 8 | 154 | 137 | |
| " | " | " | 4 | 54 | 36 | |
| Hall | " | " | 4 | 45 | 90 | |
| " | " | Carriages and Wagons | 1 | 22 | 15 | |
| " | " | Flour and Grist | 11 | 151 | 175 | |
| Milton | " | " | 4 | 68 | 82 | |
| " | " | Sawmill | 2 | 28 | 32 | |
| Lumpkin | " | " | 7 | 141 | 75 | |
| " | " | Flour and Grist | 10 | 183 | 134 | |
| " | " | Tannery | 1 | 20 | 4 | |
| " | " | Gold Mills | 3 | 35 | 700[1] | Chestatee River. |
| " | " | " | 3 | 40 | 230[1] | Vahoola Creek. |
| " | " | " | 1 | 16 | 40[1] | Cane Creek. |
| " | " | Hydraulic Mining | | 300 | 600[1] | Vahoola Ditch. |
| Habersham | " | Flour and Grist | 1 | 14 | 10 | |
| " | " | Leather | 1 | 16 | 6 | |
| " | " | Woolen-mill | 1 | 20 | 12 | |
| White | " | Flour and Grist | 1 | 14 | 15 | |
| Flint River — Campbell | " | " | 5 | 90 | 28 | |
| " — Clayton | " | " | 1 | 13 | 44 | |
| " — Fayette | " | " | 3 | 70 | 12 | |
| Tributaries of Flint River — Campbell | " | " | 8 | 148 | 50 | |
| " — Clayton | " | Sawmill | 1 | 22 | 136 | |
| " — Henry | " | Flour and Grist | 1 | 18 | 15 | |
| " — Spalding | " | " | 2 | 13 | 15 | |
| " — Fayette | " | " | 5 | 46 | 40 | |
| " — Coweta | " | " | 4 | 71 | 109 | |
| " — Meriwether | " | Sawmill | 1 | 5 | 88 | |
| " — " | " | Tannery | 1 | 30 | 12 | |
| " — " | " | Flour and Grist | 11 | 171 | 16 | |
| " — Pike | " | Sawmill | 1 | 16 | 138 | |
| " — " | " | Wheelwrighting | 1 | 8 | 15 | |
| " — " | " | Flour and Grist | 11 | 154 | 12 | |
| " — Crawford | " | " | 3 | 25 | 276 | |
| " — Upson | " | " | 15 | 191 | 43 | |
| " — " | " | Cotton | 2 | 29 | 373 | |
| | | | | | 115 | |

[1] Power estimated by B. M. Hall.

## APALACHICOLA BASIN — UTILIZED POWER — *Continued*

| STREAM | COUNTY | KIND OF MILL | No. of Mills | Total Fall Used, in Feet. | Total Net H.P. Used | REMARKS |
|---|---|---|---|---|---|---|
| Tributaries of Flint River | Upson | Sawmill | 5 | 72 | 102 | |
| " | " | Tannery | 1 | 10 | 5 | |
| " | Talbot | Flouring and Grist | 9 | 214 | 169 | |
| " | Taylor | Cotton | 1 | 12 | 40 | |
| " | Marion | Sawmill | 1 | 12 | 20 | |
| " | " | Flouring and Grist | 4 | 33 | 52 | |
| " | Taylor | " | 10 | 84 | 129 | |
| " | " | Sawmill | 6 | 58 | 95 | |
| " | Schley | Flouring and Grist | 6 | 53 | 70 | |
| " | Macon | " | 5 | 51 | 102 | |
| " | " | Sawmill | 1 | 8 | 30 | |
| " | Dooley | " | 2 | 14 | 15 | |
| " | " | Flouring and Grist | 1 | 8 | 30 | |
| " | Sumter | " | 7 | 51 | 99 | |
| " | Lee | " | 4 | 22 | 41 | |
| " | Webster | " | 8 | 66 | 107 | |
| " | " | Sawmill | 3 | 28 | 33 | |
| " | Randolph | Flouring and Grist | 6 | 69 | 84 | |
| " | Terrell | Sawmill | 2 | 11 | 30 | |
| " | " | Flouring and Grist | 2 | 14 | 15 | |
| " | Calhoun | " | 3 | 10 | 50 | |
| " | " | Sawmill | 1 | 6 | 12 | |
| " | Dougherty | Flouring and Grist | 2 | 12 | 40 | |
| " | " | Sawmill | 1 | . . | 20 | |
| " | Worth | " | 1 | 10 | 20 | |
| " | " | Flouring and Grist | 3 | 25 | 23 | |
| " | Early | Cotton | 1 | 40 | 45 | |
| " | " | Flouring and Grist | 5 | 57 | 62 | |
| " | " | Sawmill | 1 | 9 | 10 | |
| " | Miller | " | 1 | 8 | 12 | |
| " | " | Flouring and Grist | 1 | 8 | 40 | |
| " | Baker | " | 3 | 14 | 45 | |
| " | Decatur | " | 1 | 5 | 8 | |

WATER-POWERS OF GEORGIA

PLATE II

BEAN CREEK FALLS, NEAR NACOOCHEE VALLEY, WHITE COUNTY, GEORGIA.

## ALTAMAHA BASIN — IMPORTANT STREAMS

### OCMULGEE RIVER

| STREAM | TRIBUTARY TO | COUNTY | REMARKS |
|---|---|---|---|
| Ocmulgee River | Altamaha River | | |
| Mossy Creek | Indian Creek | Houston | Cotton factory; 12 ft. fall; estimated 120 H. P. (U. S. Census) |
| Indian Creek | Ocmulgee River | " | |
| Stone Creek | " | Bibb | 8 miles from Macon; 8 cu. ft. per sec.; 12 ft. fall, low water. (Locke) |
| Echaconnee Creek | " | Monroe and Crawford | Has several grist and sawmills. (U. S. Census) |
| Snake Creek | " | Twiggs and Bibb | |
| Tobesofkee Creek | " | Bibb, Monroe and Crawford | Freeman's Mill; 70 cu. ft. per sec.; 20 ft. fall, normal water. (Locke) Macon; 5 cu. ft. per sec.; 10 ft. fall, low water. (Locke) |
| Walnut Creek | " | Jones and Bibb | |
| Falling Creek | " | Jones | |
| Rum Creek | " | Monroe | |
| Towaliga River | " | Henry, Butts and Monroe | High Falls; see Power Table. Has other shoals above, and Willis Shoals nearer mouth; 10 ft. fall. Has two mills; one of them has 27 ft. head. (10th U. S. Census) |
| South Towaliga River | Towaliga River | Monroe | |
| Towaliga Creek | " | Henry | |
| Tussahaw Creek | Ocmulgee River | Henry and Butts | |
| Alcovy River | " | Newton and Walton | |
| Cornish Creek | Alcovy River | Walton | |
| Big Flat Creek | " | " | |
| Bear Creek | " | Newton | |
| South River | Ocmulgee River | " | |
| Wildcat Creek | South River | Newton | |
| Sheel Creek | " | " | |
| Walnut Creek | " | Henry | |
| Cotton River | " | " | Has several mills and sites, and is a good stream in dry weather. (10th U. S. Census) |

42　　　　　　　　　　　　　　　ALTAMAHA BASIN

## ALTAMAHA BASIN — IMPORTANT STREAMS — *Continued*

| STREAM | TRIBUTARY TO | COUNTY | REMARKS |
|---|---|---|---|
| Snap Finger Creek | South River | DeKalb | At Mitchell's mill, 20 cu. ft. per sec.; low water. (Froebel) |
| Pole Bridge Creek | " | Rockdale | 14.6 cu. ft. per sec.; extreme low water. (Froebel) |
| Honey Creek | " | " | 14.3 cu. ft. per sec.; extreme low water. (Froebel) |
| Yellow River | Ocmulgee River | Newton, Rockdale, Gwinnett | Six miles above Rockdale Paper Mill is Baker's Mill, with 9 or 10 ft. fall, and four grist-mills above it. (10th U. S. Census) |
| Big Haynes Creek | Yellow River | " | Principal tributary of Yellow River. Has many available powers, and is a fine steam in all respects. (10th U. S. Census) |
| Little Haynes Creek | Big Haynes Creek | " | |
| **OCONEE RIVER** | | | |
| Oconee River | Altamaha River | | |
| Big Sandy Creek | Oconee River | Wilkinson and Twiggs | Drainage area, 284 sq. miles. Myrick's mill, 8 ft. fall. (Locke) |
| Commissioners Creek | " | Jones and Wilkinson | Drainage area, 196 sq. miles. |
| Buffalo Creek | " | Washington | Drainage area, 286 sq. miles. |
| Palmetto Creek | " | " | Drainage area, 375 sq. miles. |
| Little River | " | Morgan and Putnam | Falls 62 ft. on five shoals in 12 miles. The largest single shoal is at Old Factory in Putnam county, 25 ft. in 900 ft. |
| Cedar Creek | Little River | Jasper, Jones and Baldwin | |
| Murder Creek | " | Jasper and Putnam | Three miles from mouth; 18 ft. fall in 600 ft. |
| Indian Creek | " | Morgan and Putnam | |
| Crooked Creek | Oconee River | Putnam | |
| Shoulderbone Creek | " | Hancock | |
| Sugar Creek | " | Morgan | |
| Apalachee River | " | Gwinnett, Walton, Oconee and Morgan | No surveys of the good powers of this river in Gwinnett and Walton counties have been made. |
| Hardlabor Creek | Apalachee River | Morgan | Has a shoal 3 miles from its mouth; 10 ft. fall. |
| Sandy Creek | Hardlabor Creek | | Has a shoal 2 miles long, 8 miles from Madison. |

## ALTAMAHA BASIN

| | | | |
|---|---|---|---|
| Shoal Creek | Apalachee River | Walton | |
| Middle Oconee River | Oconee River | Clarke, Jackson and Hall | |
| Barber's Creek | Mid. Oconee River | Oconee and Clarke | 20 ft. in 900 ft.; 24 ft. in 180 ft.; and 20 ft. in 600 ft.; all in 3 miles, near mouth; 20 ft. utilized for paper-mill. |
| Mulberry Fork | Mid. Oconee River | Jackson | Good stream for power. No surveys. |
| North Oconee River | Oconee River | Clarke, Jackson and Hall | |
| Big Sandy Creek | North Oconee River | Jackson and Clarke | |
| Walnut Fork | " | Hall | Harrington's Ford, 15.5 cu. ft. per sec.; 20 ft. fall. (Barrow) |
| Allen's Fork | " | " | County line; 22.5 cu. ft. per sec.; 10 ft. fall. (Barrow) |
| Pond Fork | " | " | Mangum's mill; 10.5 cu. ft. per sec.; 9 ft. fall. (Barrow) |
| Curry's Creek | " | Jackson | Near Jefferson; 8 cu. ft. per sec.; 18 ft. fall. (Barrow) |

## ALTAMAHA BASIN — WATER-POWERS

### Ocmulgee River

| Utilized Power | LOCATION OF POWER | POINT OF SECTION | Stage of Water | Cubic Feet per Second | Fall in Feet | Length of Shoal in Feet | Gross H.P.[1] | Source of Information | REMARKS |
|---|---|---|---|---|---|---|---|---|---|
| | **YELLOW RIVER** | | | | | | | | |
| .... | Gwinnett County | Fain's Mill | Low Spr. | 10.0 | 20.0 | .... | 136 | Barrow & Locke | Volume estimated. |
| .... | " | Steadman's Mill | " | 64.0 | 30.0 | .... | 218 | " | |
| .... | Rockdale County | Rockdale Paper-mill | Normal | 266.6 | 46.0 | 3,365 | 1,394 | B. M. Hall | |
| .... | " | Glenn Shoal | " | 283.3 | 12.0 | .... | 386 | 10th U.S. Census | |
| .... | Newton County | Bridge Shoal | " | 500.0 | 4.4 | .... | 250 | Frobel; U.S.A.E. | Volume from C. C. Anderson. |
| .... | " | Cedar Shoals | " | 515.4 | 55.0 | 2,700 | 3,221 | Anderson | Porterdale Factory, 3 m. from Covington. |
| .... | " | Dried Indian Shoal | " | 515.4 | 7.0 | 1,500 | 410 | Frobel | |
| 8 H. P. | " | Indian Fishery | LowWat'r | .... | 12.7 | 525 | 764 | Anderson | Cotton Gin. |
| | **SOUTH RIVER** | | | | | | | | |
| Utilized | DeKalb County | Flat Shoals | " | 74.0 | 24.0 | .... | 202 | Frobel | Cotton Factory of the Oglethorpe Mfg. Co. |
| .... | " | Albert Shoal | " | .... | 18.0 | .... | .... | 10th U.S. Census | Not utilized. |
| Utilized | Henry County | McKnight's Mill | " | 93.0 | 12.0 | .... | 126 | Frobel | 12 ft. head utilized; 20 ft. head available. |
| .... | " | Peachstone Shoal | " | 120.0 | 12.0 | .... | 163 | " | |
| 135 H.P. | Newton County | Snapping Shoal | Flush | 617.1 | 20.0 | 775 | .... | Anderson | 28 ft. fall in 1,500 ft. (C. C. Anderson) |
| 40 H.P. | " | Island Shoal | LowWat'r | 475.0 | 16.0 | 750 | 863 | " | |
| None | " | Mann's Bridge | " | 488.3 | 10.0 | 3,000 | 555 | " | |

[1] Net H. P. = 80% of gross H. P.

## ALTAMAHA BASIN

| H.P. | County | Shoal/Location | Low Water / Flush | | | | U.S.C. | Notes |
|---|---|---|---|---|---|---|---|---|
| | **ALCOVY RIVER** | | | | | | | |
| 30 H.P. | Newton County | White & Garner's Shoals | 55.0 | 85.0 | 3,800 | 531 | Anderson | L. W. vol. = 55 cu. ft. per sec. (10th U. S. Census) Newt'n Fc'y. Burnt during the war. |
| | **TOWALIGA RIVER** | | Flush | | | | | |
| 30 H.P. | Monroe County | High Falls | 416.6 | 85.0 | 3,800 | 4,024 | Anderson | Utilized in grist-mill. |
| | **OCMULGEE RIVER** | | LowWat'r | | | | | |
| | Newton County | Barnes' Shoals | 138.1 | 96.8 | 1,200 | 1,520 | " | |
| None | Newton County | Barnes' Shoals | " | | | 1,614 | " | At junction of South and Yellow Rivers |
| 20 H.P. | Butts County | Key's Ferry | 1,015.0 | 14.0 | 1,300 | 1,172 | " | |
| None | " | Harper or Pitman Shoal | 1,386.6 | 7.5 | 1,900 | 4,698 | " | |
| " | " | Pitman Ferry | 1,476.6 | 28.0 | 5,500 | 1,006 | " | Below ferry. |
| 20 H.P. | " | Roach's or Cargle's Shoals | 1,476.6 | 6.0 | 1,050 | 1,539 | " | At Smith's ferry. |
| Small Mill | " | Lamar's Shoals | 2,116.6 | 6.4 | 3,350 | 4,328 | " | |
| 50 H.P. | Monroe County | Glover's | 2,116.6 | 18.0 | 1,000 | 3,848 | " | |
| None | " | Dames | 2,116.6 | 16.0 | 4,000 | 1,443 | " | |
| " | " | Long or Carden's Shoals | 2,116.6 | 6.0 | 1,500 | 2,164 | " | |
| " | Bibb County | Holton | 2,116.6 | 9.0 | 4,500 | 1,449 | " | |
| " | " | Macon | 2,125.0 | 6.0 | 3,960 | | " | Fall and dist. taken from 10th U. S. Cen. |
| " | " | Proposed Macon Canal | 2,156.0 | .. | | 9,621 | " | |
| | | | 2,116.6 | 40.0 | 10 m. | | | |
| | **NORTH OCONEE RIVER** | | | | | | | |
| 32 H.P. | Jackson County | Hurricane Shoal | 76.1 | 30.0 | 600 | 237 | | |
| .. H.P. | " | Tumbling Shoal | 126.2 | 8.0 | 600 | 113 | | |
| 200 | Clarke County | Athens Factory | 331.9 | 12.0 | | | | At Athens. |
| 200 | " | Georgia Factory | .. | 21.0 | 2,100 | 704 | Anderson | Near junction of rivers. |
| .. | Hall County | Carnesville and Gainesville Road | 31.5 | 10.0 | .. | 34 | Barrow | |
| | **MIDDLE OCONEE RIVER** | | LowWat'r | | | | | |
| None | Jackson County | Tallasee Bridge | 241.3 | 32.0 | 3,600 | 999 | Anderson | Total fall said to be 58 ft. in less than a mile. |
| .. H.P. | Clarke County | Mc'Elroy's Mill | 241.3 | 23.0 | 2,600 | 718 | " | |
| 60 H.P. | " | Princeton Factory | 290.6 | 15.0 | Dam | 495 | " | |

ALTAMAHA BASIN — WATER-POWERS — *Continued*

| Utilized Power | LOCATION OF POWER | POINT OF SECTION | Stage of Water | Cubic Feet per Second | Fall in Feet | Length of Shoal in Feet | Gross H. P.[1] | Source of Information | REMARKS |
|---|---|---|---|---|---|---|---|---|---|
| | APALACHEE RIVER | | | | | | | | |
| | Oconee County | Just above High Shoals | | | 20.0 | | | U. S. Cen. | |
| 150 H. P. | " " | High Shoals | Normal | 139.6 | 50.0 | 600 | 792 | Anderson | |
| 30 H. P. | " " | Price's Mill | " | 139.6 | 19.0 | 900 | 301 | " | |
| | Morgan County | Furlow's Shoals | Low Wat'r | 47.0 | 26.0 | 4,200 | 139 | U. S. Cen. | 8' at mill, and 18' above. |
| | " " | Reid's Mill | " | 76.0 | 8.0 | | 69 | " | |
| | OCONEE RIVER | | | | | | | | |
| 150 H. P. | Oconee County | Barnett's Shoal | " | 624.1 | 54.0 | 4,000 | 3,830 | Anderson | 5 miles below junction of Middle and North Oconee rivers. |
| | Morgan County | Scull's Shoal | " | | 10.0 | Dam | | 10th U.S. Census | Powell Mfg. Co.'s dam backs water 2 miles. |
| | " " | Park's Mill | " | | 8.0 | " | | " | Grist-mill. |
| | Intervening two Shoals | | " | | 7.0 | " | | " | |
| | Putnam County | Long Shoal | " | 533.3 | 12.0 | 1,300 | 726 | " | Old factory site, not in use. Head can be made 15 or 20 feet by dam. |
| | Intervening six Shoals | | " | | 33.0 | | | | |
| | Baldwin County | Milledgeville | " | 740.0 | 34.0 | 5 or 6 m. | 2,859 | 10th U.S. Census | Canal proposed. |
| | Hall County | Six miles from Gainesville | " | 30.0 | 39.0 | | 133 | Anderson | Head-waters. |
| | LITTLE RIVER | | | | | | | | |
| | Putnam County | Site of old Eatonton Fact'ry | Low Wat'r | 45.0 | 25.0 | 900 | 127 | 10th U.S. Census | Volume estimated. No utilized power. |
| | " " | Grist Mill | " | | 8.0 | | | " | |
| | " " | Pierson's Mill | " | | 13.5 | | | " | |
| | " " | " | " | | 7.0 | | | " | |
| | " " | Humber's Mill | " | 108.0 | 9.0 | | 110 | " | Volume estimated. |

[1] Net H. P. = 80 per cent. of gross H. P.

NOTE.— The foregoing is a very imperfect statement, concerning the water-powers of the Altamaha Basin; but it is the best that can be done with the data at hand.

## ALTAMAHA BASIN — UTILIZED POWER

| STREAM | COUNTY | KIND OF MILL | No. of Mills | Total Fall Used, in Feet | Total Net H.P. Used | REMARKS |
|---|---|---|---|---|---|---|
| Tributaries to Altamaha R. | Tattnall | Flour and Grist | 3 | | 62 | |
| " | Johnson | Sawmill | 2 | 21 | 55 | |
| " | Baldwin | Flour and Grist | 2 | 15 | 24 | |
| " | Putnam | " " " | 2 | 12 | 70 | |
| Oconee River | Greene | " " " | 2 | 15 | 70 | |
| " | " | Cotton Factory | 1 | 10 | | |
| " | " | Flour and Grist | 3 | 26 | 104 | |
| " | Clarke | " " " | 1 | 8 | 6 | |
| " | Putnam | " " " | 4 | 32 | 165 | |
| Little River | " | Sawmill | 1 | 7 | 20 | |
| " | Morgan | Flour and Grist | 2 | 22 | 25 | |
| " | Newton | " " " | 2 | 47 | 30 | |
| " | Walton | Cotton Gin | 1 | 25 | 15 | |
| " | Morgan | Flour and Grist | 1 | 40 | 45 | |
| Apalachee River | Walton | " " " | 1 | 20 | 20 | |
| " | " | Cotton Factory | 1 | 20 | 100 | |
| " | Gwinnett | Flour and Grist | 5 | 42 | 124 | |
| " | " | " " " | 1 | 22 | 10 | |
| Other Tributaries of Oconee River | Laurens | " " " | 3 | 34 | 50 | |
| " | Johnson | Sawmill | 2 | 22 | 50 | |
| " | Twiggs | Flour and Grist | 2 | 16 | 23 | |
| " | " | " " " | 3 | | 63 | |
| " | Washington | Sawmill | 1 | 6 | 20 | |
| " | " | Flour and Grist | 3 | | 58 | |
| " | Wilkinson | " " " | 12 | 69 | 140 | |
| " | " | Sawmill | 8 | 4 | 102 | |
| " | " | Agricultural Implements | 1 | 3 | 4 | |
| " | Hancock | Flour and Grist | 6 | 94 | 95 | |
| " | Jones | " " " | 4 | 60 | 98 | |
| " | Baldwin | " " " | 3 | 37 | 60 | |
| " | Jasper | " " " | 2 | 30 | 32 | |

ALTAMAHA BASIN — UTILIZED POWER — *Continued*

| STREAM | COUNTY | KIND OF MILL | No. of Mills | Total Fall Used, in Feet | Total Net H.P. Used | REMARKS |
|---|---|---|---|---|---|---|
| Other Tributaries of Oconee River | Putnam | Flour and Grist | 6 | 73 | 178 | |
| " | Morgan | Sawmill | 1 | 8 | 25 | |
| " | Walton | Flour and Grist | 7 | | 90 | |
| " | Greene | " " " | 6 | 91 | 122 | |
| " | " | Sawmill | 1 | 16 | 50 | |
| " | Oconee | Cotton Gin | 2 | 23 | 32 | |
| " | Oglethorpe | Flour and Grist | 1 | 41 | 11 | |
| " | " | " " " | 2 | 22 | 30 | |
| " | " | Sawmill | 4 | 56 | 30 | |
| North Oconee River | Gwinnett | Woolen-mill | 1 | 128 | 100 | |
| Middle Oconee River | Clarke | Cotton Factory | 2 | 16 | 12 | |
| North and Middle Oconee and Tributaries | " | " | 1 | 32 | 330 | |
| " | " | Sawmill | 1 | 20 | 100 | |
| " | " | Paper-mill | 1 | 12 | 10 | |
| " | Gwinnett | Flour and Grist | 4 | 16 | 75 | |
| " | " | Sawmill | 1 | 52 | 82 | |
| " | Madison | Flour and Grist | 2 | 32 | 26 | |
| " | Hall | " " " | 11 | 12 | 12 | |
| " | " | Sawmill | 8 | 29 | 13 | |
| " | Jackson | " | 13 | 170 | 130 | |
| " | " | Flour and Grist | 5 | 16 | 15 | |
| " | " | Cotton Gin | 1 | 146 | 141 | |
| " | " | Leather | 1 | 201 | 187 | |
| " | " | Woolen-mill | 1 | 82 | 70 | |
| Ocmulgee River | Monroe | Flour and Grist | 1 | 30 | 10 | |
| " | Jones | " " " | 4 | 8 | 6 | |
| " | " | " " " | 1 | 12 | | |
| " | Butts | Sawmill | 1 | 12 | | |
| | | | | 48 | 103 | |
| | | | | 12 | 40 | |

THE NATURAL DAM, BIG POTATO CREEK, NEAR THOMASTON, UPSON COUNTY, GEORGIA.

## ALTAMAHA BASIN

| River | County | Type | | | |
|---|---|---|---|---|---|
| Ocmulgee River | Jasper | Woolen-mill | 1 | 12 | 6 |
| " " | Henry | Flour and Grist | 2 | 34 | 14 |
| Tributaries of Ocmulgee R. | Wilcox | Sawmill | 1 | 6 | 4 |
| " " | Wilcox | " | 1 | 6 | 24 |
| " " | Dodge | Flour and Grist | 1 | | 10 |
| " " | Pulaski | " | 5 | 45 | 46 |
| " " | " | Woolen-mill | 1 | 9 | 4 |
| " " | Houston | Sawmill | 1 | 9 | 15 |
| " " | " | " | 3 | 25 | 46 |
| " " | " | Flour and Grist | 10 | | 186 |
| " " | Twiggs | Cotton Factory | 1 | 12 | 60 |
| " " | Crawford | Flour and Grist | 1 | 8 | 11 |
| " " | Bibb | " " | 3 | 36 | 90 |
| " " | " | Sawmill | 1 | 9 | 20 |
| " " | " | Cotton Gin | 1 | 13 | 30 |
| Towaliga River | Monroe | Sawmill | 1 | | 8 |
| " " | " | " | 1 | 11 | 12 |
| " " | " | Flour and Grist | 3 | 39 | 15 |
| " " | " | Wool Carder | 1 | 5 | 76 |
| " " | Henry | Flour and Grist | 2 | 100 | 4 |
| Alcovy River | Newton | Sawmill | 2 | 30 | 120 |
| " " | " | Cotton Gin | 1 | 6 | 36 |
| " " | " | Flour and Grist | 2 | 30 | 20 |
| " " | Walton | Sawmill | 1 | 19 | 40 |
| " " | Gwinnett | Flour and Grist | 2 | 66 | 15 |
| Yellow River | Newton | Wheelwright | 3 | 34 | 18 |
| " " | " | Cotton Factory | 1 | 14 | 54 |
| " " | " | Paper-mill | 1 | 16 | 5 |
| " " | " | Flour and Grist | 1 | 20 | 76 |
| " " | " | Sawmill | 2 | 21 | 60 |
| " " | Rockdale | Flour and Grist | 2 | 24 | 25 |
| " " | " | Sawmill | 1 | | 80 |
| " " | Rockdale | Cotton Gin | 1 | 14 | 70 |
| " " | " | Furniture | 1 | 14 | 10 |
| " " | " | Paper-mill | 1 | 14 | 10 |
| " " | DeKalb | Flour and Grist | 1 | 18 | 10 |
| " " | " | Cotton Gin | 1 | 7 | 90 |
| " " | " | " | 1 | 7 | 15 |
| " " | " | | | | 6 Rockdale Paper-mill |

## ALTAMAHA BASIN — UTILIZED POWER — Continued

| STREAM | COUNTY | KIND OF MILL | No. of Mills | Total Fall Used, in Feet | Total Net H.P. Used | REMARKS |
|---|---|---|---|---|---|---|
| Yellow River | Gwinnett | Flour and Grist | 6 | 66 | 126 | |
| " | " | Furniture | 1 | 8 | 10 | |
| " | " | Sawmill | 1 | 14 | 15 | |
| South River | DeKalb | Cotton Factory | 1 | 23 | .. | |
| " | Henry | Flour and Grist | 1 | 8 | 20 | |
| " | " | Agricultural Implements | 1 | 9 | 3 | |
| " | " | Furniture | 1 | 9 | 3 | |
| " | " | Sawmill | 1 | 9 | 20 | |
| " | Newton | Flour and Grist | 1 | 30 | 10 | |
| " | " | " | 2 | 24 | 25 | |
| " | Rockdale | Cotton Gin | 1 | 16 | 39 | |
| " | " | Furniture | 1 | 9 | 4 | |
| " | DeKalb | Flour and Grist | 2 | 35 | 65 | |
| " | " | Sawmill | 1 | 10 | 15 | |
| " | " | Cotton Gin | 1 | 10 | 12 | |
| " | " | Furniture | 1 | 10 | 5 | |
| " | Fulton | Sawmill | 1 | 22 | 9 | |
| " | " | Flour and Grist | 2 | 34 | 24 | |
| Other Trib's of Ocmulgee R. | Pike | " | 2 | 74 | 55 | |
| " | Monroe | " | 11 | 157 | 148 | |
| " | " | Sawmill | 1 | 11 | 9 | |
| " | " | Cotton Gin | 1 | 11 | 5 | |
| " | Henry | Flour and Grist | 3 | 78 | 38 | |
| " | " | Sawmill | 2 | 33 | 23 | |
| " | Butts | Flour and Grist | 4 | 52 | 45 | |
| Tributaries of South River | Henry | " | 3 | 109 | 26 | |
| " | " | Sawmill | 1 | 10 | 10 | |
| " | " | Woolen mill | 1 | .. | 5 | |
| " | Clayton | Flour and Grist | 2 | 36 | 33 | |
| " | Rockdale | " | 3 | 62 | 48 | |
| " | " | Sawmill | 1 | 18 | 6 | |

## ALTAMAHA BASIN

| | County | Industry | No. | | |
|---|---|---|---|---|---|
| | Rockdale | Cotton Gin | 2 | 31 | 22 |
| | " | Leather | 1 | 8 | 4 |
| | " | Flour and Grist | 1 | 30 | 12 |
| | Newton | " | 10 | 180 | 128 |
| | DeKalb | Sawmill | 3 | 44 | 30 |
| | " | Cotton Gin | 6 | 108 | 54 |
| | " | Paper-mill | 3 | 99 | 152 |
| | " | Leather | 1 | 15 | 20 |
| Tributaries of Yellow River | Newton | Cotton Gin | 2 | 15 | 15 |
| | " | Flour and Grist | 2 | 37 | 18 |
| | " | Cotton Gin | 1 | 12 | 8 |
| | Rockdale | Flour and Grist | 3 | 70 | 73 |
| | " | Sawmill | 1 | . | 13 |
| | Walton | Flour and Grist | 1 | 15 | 8 |
| | " | " | 3 | 35 | 22 |
| | Gwinnett | " | 2 | 51 | 10 |
| | DeKalb | " | 2 | 26 | 25 |
| | " | Sawmill | 2 | 55 | 20 |
| | " | Cotton Gin | 2 | 32 | 33 |
| | " | Furniture | 1 | 15 | 3 |
| | " | Flour and Grist | 1 | 18 | 8 |
| Tributaries of Alcovy River | Walton | " | 2 | 54 | 32 |
| | " | Cotton Gin | 1 | 15 | 5 |
| | Gwinnett | Sawmill | 1 | 18 | 20 |

# THE OGEECHEE BASIN

The greater part of this drainage basin lies below the fall-line, and, as only that part, which lies above the fall-line, has much importance for water-power, this is the smallest and least important of the six basins, considered from the standpoint of water-power. The first power, in going up the Ogeechee river, is at the fall-line, and is known as the *Shoals of the Ogeechee*. They are above the mouth of Little Ogeechee, 8½ miles from Mayfield, the nearest railroad station. Part of the power is utilized by a grist- and saw-mill. The entire fall of the shoal is 21 feet; but the mill utilizes only about 18 feet, and about 40 net, 12-hour horse-power. The low season volume is estimated at 40 cubic-feet per second. With the fall, head and storage, 200 gross, 12-hour horse-power is available. *The Jewell Cotton Factory*, 4½ miles from Mayfield, is the next power. For eight months in the year, 150 net, 12-hour horse-power is utilized with storage. During the other four months, it is sometimes necessary to use auxiliary steam-power to the extent of 125 horse-power.

Nearly all the power on this basin being utilized, the following tabulated statement from the 10th U. S. Census is given, as the best showing, that can be made. It is the only data available.

## OGEECHEE BASIN — UTILIZED POWER

| STREAM | COUNTY | KIND OF MILL | No. of Mills | Fall Used, in Feet | Total Net H.P. Used | REMARKS |
|---|---|---|---|---|---|---|
| Ogeechee River | Warren | Flour and Grist-mill | 2 | 20.0 | 30 | |
| " | Hancock | " | 2 | 13.0 | 40 | |
| " | " | Woolen Mill (Carder) | 1 | | 8 | |
| " | Warren | Cotton Factory | 1 | 16.0 | 150 | |
| " | Taliaferro | Flour and Grist-mill | 1 | 22.0 | 15 | |
| Tributaries to Ogeechee River | Liberty | Sawmill | 1 | 9.0 | 20 | |
| " | Bulloch | " | 2 | | 27 | |
| " | " | Flour and Grist-mill | 5 | 36.0 | 20 | |
| " | " | Sawmills | 2 | 17.5 | 24 | |
| " | Screven | Flour and Grist-mill | 1 | 10.0 | 8 | |
| " | " | Sawmill | 1 | 10.0 | 12 | |
| " | Burke | Flour and Grist-mill | 9 | 75.0 | 117 | |
| " | Jefferson | " | 9 | 82.0 | 189 | |
| " | Washington | " | 4 | 21.0 | 33 | |
| " | Glascock | " | 2 | 60.0 | 54 | |
| " | Hancock | Sawmill | 2 | 23.0 | 27 | |
| " | " | Flour and Grist-mill | 2 | 42.0 | 30 | |
| " | Warren | " | 1 | 9.0 | 12 | |

## SAVANNAH BASIN — IMPORTANT STREAMS

| STREAM | TRIBUTARY TO | COUNTY | REMARKS |
|---|---|---|---|
| Savannah River | Atlantic Ocean | | Jacksonboro, 87.3 cu. ft. per sec.; 7 ft. fall. (Barrow) |
| Beaverdam Creek | Savannah River | Screven | Mill Haven, 565.5 cu. ft. per sec.; 10 ft. fall. (Barrow) |
| Briar Creek | " | " | Wade's Mill, 12 cu. ft. per sec.; 5 ft. fall. (Barrow) |
| Rocky Creek | " | " | 12 cu. ft. per sec.; 8 ft. fall. (Barrow) Little |
| Spirit Creek | " | Richmond | Spring Cr. at mouth. |
| Butler's Creek | " | " | |
| Rock Creek | " | " | |
| Bottle's Creek | " | Columbia | |
| Kioker Creek | " | " | Near Appling, 30 cu. ft. per sec.; 10 ft. fall, low water. (Barrow) |
| Keg Creek | " | " | |
| Little River | " | McDuffie | Power at Mrs. J. Belknap Smith's, 47 cu. ft. per sec.; 8 ft. fall; 218 H. P. utilized by six mills on river. |
| Sweetwater Creek | " | " | Cotton card factory; 21 ft. head; 36 gross H. P. (Barrow) |
| Soap Creek | " | Lincoln | |
| Fishing Creek | " | " and Wilkes | |
| Pistol Creek | " | " | |
| Broad River | " | Franklin, Madison, Oglethorpe etc. | Franklin Co., Toccoa and Carnesville Road, 50 cu. ft. per sec.; low spring. (Barrow) |
| Long Creek | Broad River | Oglethorpe | 4 m. from Lexington, 7.2 cu. ft. per sec.; 10 ft. fall. (Barrow) |
| S. Fork, Broad River | " | " | At Eberhart's Mill, 80 ft. fall in 1 m. (U. S. Cens.) At Watson's Mill, 30 ft. fall in 1 m. (U. S. Cens.) |
| Groves Creek | S. Fork, Broad River | " | |
| Cloud's Creek | " | " | |
| Beaverdam Creek | " | " and Madison | |
| Millshoal Creek | " | Madison | |
| Bushy Creek | " | Franklin and Madison | |
| N. Fork, Broad River | Broad River | | |

## SAVANNAH BASIN

| | | | |
|---|---|---|---|
| Hudson's Fork | N. Fork, Broad River | Banks and Franklin | { Homer and Mt. Airy Road, 77.3 cu. ft. per sec., normal. (Locke) <br> { 4 miles from Carnesville, 50 cu. ft. per second, normal. (Barrow) |
| Unawattee Creek | " " | Franklin | |
| Webb's Creek | Hudson Fork, Br'd R. | Banks | |
| Bear Creek | N. Fork, Broad River | Franklin | |
| Beaverdam Creek | Savannah River | Elbert | { Point east of Southern R'y, 30 cu. ft. per sec. (Barrow) <br> { Stream has 9 mills and several good undeveloped shoals. (U. S. C.) |
| Cold Water Creek | " " | Elbert | |
| Lightwood Log Creek | " " | Hart | |
| Tugalo River | " " | | |
| Panther Creek | Tugalo River | Habersham | Walker's mill, 4.5 cu. ft. per sec.; 20 ft. fall. (Barrow) |
| Tallulah River | " " | Rabun | Tallulah Falls. (See Power Table) |
| Toccoa Creek | " " | | { Toccoa Falls, 5.2 cu. ft. per sec.; 190 ft. fall. (Barrow & Locke) <br> { Parker's mill, 333.7 cu. ft. per sec., normal. (C. C. Anderson) |
| Persimmon Creek | Tallulah River | Rabun | |
| Chatuga River | Tugalo River | Rabun | { Near Clayton, 3.7 cu. ft. per sec. At mouth, 30 cu. ft. per sec. (Barrow) |
| Stekoa Creek | Chatuga River | Rabun | |
| War Woman Creek | " | Rabun | |
| Wildcat Creek | " | Rabun | At mouth, 50 cu. ft. per sec., low water. (Barrow) |
| Tiger Creek | " | Rabun | At mouth, 40.6 cu. ft. per sec., low water. (Barrow) |

## SAVANNAH BASIN — WATER-POWERS

| LOCATION OF WATER-POWER | POINT OF SECTION | Stage of Water | Cubic Feet per Second | Fall in Feet | Length of Shoal, in Feet | Gross H.P.[1] | Source of Information | REMARKS |
|---|---|---|---|---|---|---|---|---|
| TALLULAH RIVER | | | | | | | | |
| Rabun County | Tallulah Falls | Normal | 723.3 | 335.0 | 4,000 | 27,470 | Anderson | |
| TUGALO RIVER | | | | | | | | |
| Habersham County | Mouth of Tallulah River | Low Wat'r | 654.0 | 75.0 | 2½ m. | 5,573 | J. P. Carson, Ass't U.S. Eng. | |
| Franklin County | Eastonolly Shoals | " | .... | 4.0 | 2,640 | .... | " | |
| " | Stribling Shoals | " | .... | 2.0 | 2,640 | .... | " | |
| Hart County | Guest Shoal | " | 290.0 | 17.0 | 5,280 | 560 | 10th U. S. Census | |
| " | Hatton Shoal | " | 290.0 | 39.0 | 8,000 | 1,280 | " | |
| BROAD RIVER | | | | | | | | |
| Elbert County | Baker's Ferry | " | 600.0 | 3.0 | 600 | 204 | " | Fall said to be over 70 ft. in 1¼ miles. (U. S. Census) |
| " | Anthony's Shoals | " | 600.0 | 70.0 | 6,600 | 4,772 | " | |
| " | Smith Shoals | " | 600.0 | 10.0 | 2,640 | 681 | " | |
| SAVANNAH RIVER | | | | | | | | |
| Hart County | McDaniel's Shoals | .. | 766.6 | 30.0 | 5 m. | 2,600 | " | Volume as given by U. S. Eng. J. P. Carson, 1,275 cu. ft. per second. |
| Elbert County | Ferrill's Ledge | .. | 766.6 | 3.0 | 360 | 260 | " | Vol. etc., 1,750 cu. ft. per sec. |
| " | Middleton's Shoal | .. | 833.3 | 18.0 | 5,280 | 1,700 | " | Vol. etc., 1,873.3 " " |
| " | Gregg's Shoal | .. | 833.3 | 14.0 | 5,280 | 1,325 | " | Vol. etc., 2,000 " " |
| " | Bowman's Ledge | .. | 880.0 | 3.0 | 120 | 300 | " | Vol. etc., 2,100 " " |
| " | Cherokee Shoal | .. | 880.0 | 9.0 | 2,640 | 900 | " | Vol. etc., 2,150 " " |
| Lincoln County | Trotter's Shoal | .. | 107.5 | 75.0 | 7 m. | 9,165 | " | Vol. etc., 2,400 " " |
| Columbia County | Long Shoal | .. | 1,800.0 | 35.0 | 5 m. | 7,250 | " | Vol. etc., 2,775 " " |
| Richmond County | Blue Jacket Shoal | .. | 2,166.6 | 10.0 | 600 | 2,350 | " | |
| " | Augusta | L. Season Dry Y'rs Max. with Storage | 2,400.0 | 50.0 | Canal 7 miles | 13,636 | " | The city owns the water-power and factory sites, Mfg. Cos. buy sites, and lease power. |
| " | Same with average head attainable | L. Season Dry Y'rs | 6,000.0 2,400.0 | 50.0 40.0 | " " | 34,090 10,908 | " " | |

[1] Net H. P. = 80 per cent. of gross H. P.

FLAT SHOALS ON THE FLINT RIVER, BETWEEN PIKE AND MERIWETHER COUNTIES, GEORGIA.

## SAVANNAH BASIN — UTILIZED POWER

| STREAM | COUNTY | KIND OF MILL | No. of Mills | Total Fall Used | Total Net H.P. Used | REMARKS |
|---|---|---|---|---|---|---|
| Savannah River | Richmond | Miscellaneous | 15 | .... | 3,650 | |
| " | Lincoln | Flour and Grist | 3 | 14 | 32 | |
| " | Elbert | " | 2 | 19 | 115 | |
| Tributaries of Savannah R. | Effingham | Sawmill | 1 | 6 | 20 | |
| " | Burke | Flour and Grist | 8 | 72 | 96 | |
| " | Richmond | " | 11 | 125 | 190 | |
| " | " | Sawmill | 8 | 100 | 209 | |
| " | " | Cotton Factory | 1 | 9 | 50 | |
| " | " | Woolen-mill | 1 | 9 | 45 | |
| Little River | Lincoln | Sawmill | 3 | 24 | 45 | |
| " | " | Flour and Grist | 4 | 30 | 60 | |
| " | McDuffie | Gold Stamp-mill | 1 | 9 | 60 | |
| " | Wilkes | Flour and Grist | 1 | 8 | 12 | |
| " | Warren | " | 1 | 8 | 8 | |
| " | Greene | Saw and Grist | 1 | 8 | 30 | |
| Other Tributaries of Savannah River | Columbia | Flour and Grist | 5 | 14 | 5 | |
| " | McDuffie | Sawmill | 1 | 69 | 91 | |
| " | " | Flour and Grist | 7 | 10 | 25 | |
| " | Warren | " | 1 | 127 | 152 | |
| Broad River and Tributaries | Oglethorpe | Sawmill | 10 | 20 | 15 | |
| " | Madison | Flour and Grist | 10 | 12 | 12 | |
| " | " | " | 5 | 195 | 175 | |
| " | Elbert | Sawmill | 3 | 145 | 281 | |
| " | Franklin | Flour and Grist | 9 | 61 | 64 | |
| " | " | Cotton Gin | 4 | 44 | 39 | |
| " | Banks | Sawmill | 6 | 56 | 163 | |
| Other Tributaries of Savannah River | Wilkes | Flour and Grist | 1 | 83 | 54 | |
| " | Elbert | " | 12 | 18 | 53 | |
| | | Flour and Grist | 7 | 169 | 20 | |
| | | " | 6 | 85 | 279 | |
| | | | | 73 | 75 | |
| | | | | | 134 | |

## SAVANNAH BASIN — UTILIZED POWER — Continued

| STREAM | COUNTY | KIND OF MILL | No. of Mills | Total Fall Used | Total Net H.P. Used | REMARKS |
|---|---|---|---|---|---|---|
| Other Tributaries to Savannah River | Elbert | Sawmill | 1 | 14 | 12 | |
| " | Hart | Flour and Grist | 11 | 194 | 156 | |
| " | " | Sawmill | 1 | 14 | 15 | |
| " | " | Cotton Gin | 8 | 99 | 50 | |
| Tributaries of Tugalo River | " | Sawmill | 1 | 30 | 10 | |
| " | " | Flour and Grist | 2 | 27 | 45 | |
| " | " | Cotton Factory | 1 | 26 | 20 | |
| " | " | Wool Carder | 1 | 20 | 44 | |
| " | Habersham | Flour and Grist | 4 | 47 | 46 | |
| " | " | Leather | 1 | 16 | 6 | |
| " | " | Sawmill | 3 | 46 | 58 | |
| " | " | Woolen-mill | 1 | . . | 6 | |
| " | Rabun | Sawmill | 1 | 14 | 8 | |

## OCKLOCKONEE AND SUWANNEE BASINS — UTILIZED POWER

| STREAM | COUNTY | KIND OF MILL | No. of Mills | Total Fall Used | Total Net H.P. Used | REMARKS |
|---|---|---|---|---|---|---|
| Ocklockonee R. and Trib'r's | Colquitt | Flour and Grist | 3 | 16 | 30 | |
| " | Decatur | " | 4 | 64 | 50 | |
| " | " | Sawmill | 1 | 6 | 12 | |
| " | Thomas | Flour and Grist | 4 | 32 | 34 | |
| Ocilla R. and Tributaries | " | " | 4 | 60 | 50 | |
| Tributaries of the Suwannee River | Berrien | Woolen-mill | 1 | 12 | 12 | |
| " | " | Flour and Grist | 10 | 82 | 145 | |
| " | " | Woolen-mill | 1 | 9 | 10 | |
| " | Brooks | Sawmill | 1 | . . | 12 | |
| " | " | Flour and Grist | 7 | 43 | 54 | |
| " | Clinch | " | 1 | 7 | 15 | |
| " | Echols | " | 1 | 6 | 6 | |
| " | " | Cotton Gin | 1 | 12 | 6 | |
| " | Lowndes | Sawmill | 1 | 10 | 10 | |
| " | " | Flour and Grist | 8 | 80 | 77 | |
| " | Wilcox | " | 1 | 6 | 4 | |

# CHAPTER IV

## FLOW OF STREAMS

The object of this chapter is to show, in a concise manner, the important facts, developed by the water-power surveys of Mr. C. C. Anderson, C.E., late Assistant State Geologist.

The new and special feature, presented, is a compilation from his notes, showing in tabulated form the daily fluctuation, for thirteen consecutive months, at certain points on the Chattahoochee, the Flint and the Ocmulgee rivers, each table being accompanied by a cross-section of the stream, and by velocities taken with a Haskell current-meter at certain stages. From this, discharges, in cubic feet per second, are calculated. This is the first systematic attempt, at gauging any of the streams of the State, to determine their flow, at all seasons of the year. Unfortunately, it covers a very limited portion of the wide field, that is open for investigation; but the results are very gratifying, as far as they go. They make a good showing for the constancy of these streams, and will be of incalculable value to the hydraulic engineer, in future investigations.

The important items, that determine the value of any water-power, are: — *First*, the quantity of water flowing in the stream, at all seasons of the year; *second*, the available fall; *third*, its location; *fourth*, the cost of development. A competent engineer can determine the last three of these items, in a short time, at any season of the year; but the first cannot be determined, in a short time. It must be found by a series of gaugings, extending over at least twelve consecutive months, and, preferably, a great deal longer time. In the absence of data, obtained in this way, engineers are forced to form an estimate from the area and the character of the water-shed, rainfall, statistics etc. A short method, frequently adopted, and which often leads to glaring errors, is to figure out a low water-flow for the river, at so many cubic feet per minute, for each square mile of water-shed, using, as a standard, the measured low water-flow of some other stream, assumed to be identical in its characteristics. But the water-shed rule, which applies to one stream, cannot be applied at random to all other streams, which seem to have the same general character of water-shed. Each stream has its own peculiarities; and, while it is a comparatively easy matter to arrive at an estimate of the total annual discharge, or run-off, and form a tolerably correct idea of the amount of water available for storage, when the form and area of the water-shed, geological formations and rainfall are known, the data, as to the low water-flow of a stream, must be derived from the actual daily fluctuations and measurements of discharge at known stages. When enough data of this kind has accumulated, a reliable curve of discharge can be made. In Mr. Anderson's work, the velocity was not metered, often enough, to give a complete curve of discharge; but some of the meterings were taken at such low stages, that a close approximation to the minimum, for the period covered by his observations, can be arrived at. Mr. Anderson established gauge-stations on the Chattahoochee, Flint and Ocmulgee rivers, in August, 1891. At each station, a gauge-rod, divided into feet and tenths of feet, was set vertically in the stream, and firmly attached to a bridge-pier or some other permanent object. The rod was made of sufficient length, to cover the fluctuations of the stream, and its bottom end placed low enough in the water to be below the surface, at lowest stage. A gauge-reader, residing in the vicinity, was then employed, whose duty it was, to read the surface height of the water, every morning, and keep a record of it. Some of these gauge-readers failed to note the stage of water, for several days at a time, thus causing blanks in the tables here presented; but, where these blanks occur, there are generally other stations on the same stream, that give the reading, for that day, and thus show, comparatively, the stage of water.[1]

The following fluctuation-tables are made from these records. The readings, which were elevations above the bottom of the gauge-rod, have been reduced to elevations above the lowest observed water, which is the "0.0" of the table, when given.

At each gauge-station, a cross-section of the stream was made, as shown in connection with the tables. The cross-section was divided into subsections, from five to fifty feet in width; and the velocity, in feet per second, was taken at each subsection, with a Haskell current-meter, the stage of the stream being noted from the gauge-rod, at the time. At most of the stations, there has been a metering of the streams, at a low stage of water, so near to the minimum observed stage, that the velocities, $v$ and $v'$, at given stage, and minimum observed stage, would be approximately proportional to the square roots of the respective areas of water-way, $a$ and $a'$. So that, $v' = v \sqrt{\dfrac{a'}{a}}$. In this way, the compiler has calculated a volume for minimum observed stage, at four of these stations, based on Mr. Anderson's lowest actual measurements. At two others, Mr. Anderson's statement, concerning the volume at minimum observed stage, has been given.

Gauge-stations were also established on four branches of the Oconee river in January, 1893; but the only gauge-readings were for January and February, which was too short a time to render them of any value.

---

[1] Compare tables for Porter Mills, Roswell, West Point and Columbus.

FIG. 1

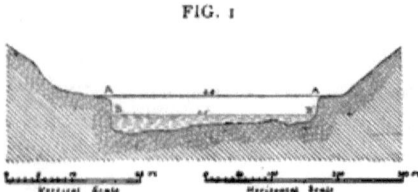

CROSS-SECTION OF THE SOQUEE RIVER, AT PORTER MILLS, HABERSHAM COUNTY, GEORGIA

## FLOW OF THE SOQUEE RIVER, AT PORTER MILLS, HABERSHAM COUNTY, GEORGIA

| No. | Date of Measurement | Stage | Area in Sq. Ft. of Cross-Section | Velocity in Feet per Second | Discharge in Cu Ft. per Sec. | REMARKS |
|---|---|---|---|---|---|---|
| 1 | Dec. 1890 | Not given | 306.6 | 0.90 | 275.9 | Different section, measured by C. C. Anderson. |
| 2 | Aug. 13, 1891 | 0.6 | 336.0 | 1.22 | 409.9 | Section, here given, was measured by C. C. Anderson. |
| 3 | . . . . . . | 0.0 | 250.0 | 1.66 | 266.6 | Section calculated. |

## TABLE I
### DAILY FLUCTUATIONS IN FEET AND TENTHS
*Lowest Observed Stage = 0.0*

THE SOQUEE RIVER AT PORTER MILLS, HABERSHAM COUNTY, GEORGIA

| Date | 1891 | | | | | 1892 | | | | | | | | |
|---|---|---|---|---|---|---|---|---|---|---|---|---|---|---|
| | Aug. | Sept. | Oct. | Nov. | Dec. | Jan. | Feb. | Mar. | Apr. | May | June | July | Aug. | Sept. |
| 1 |  | 0.7 | 0.2 | 0.2 | 0.3 | 0.3 | 0.5 | 0.5 | 0.5 | 0.8 | 0.5 | 0.7 | 0.6 | 0.5 |
| 2 |  | 0.4 | 0.2 | 0.2 | 0.3 | 0.3 | 0.5 | 0.5 | 0.5 | 0.8 | 0.8 | 0.7 | 0.6 | 0.5 |
| 3 | Begun August 13th, 1891. | 0.4 | 0.2 | 0.2 | 0.3 | 0.7 | 0.5 | 0.5 | 0.5 | 0.8 | 1.0 | 0.7 | 0.5 | 0.5 |
| 4 | | 0.5 | 0.2 | 0.2 | 0.3 | 0.4 | 0.5 | 0.5 | 0.5 | 0.8 | 1.3 | 0.7 | 0.5 | 0.5 |
| 5 | | 0.2 | 0.2 | 0.2 | 0.4 | 0.5 | 0.5 | 0.5 | 0.5 | 0.8 | 0.8 | 1.8 | 0.5 | 0.5 |
| 6 | | 0.3 | 0.2 | 0.2 | 0.6 | 0.7 | 0.5 | 0.6 | 2.8 | 0.7 | 0.8 | 1.2 | 0.5 | 0.5 |
| 7 | | 0.3 | 0.3 | 0.2 | 0.2 | 0.5 | 0.5 | 0.7 | 2.0 | 0.7 | 0.8 | 1.0 | 0.5 | 0.5 |
| 8 | | 0.3 | 0.2 | 0.5 | 0.0 | 0.5 | 0.7 | 1.2 | 1.7 | 0.7 | 0.8 | 0.8 | 1.0 | 0.5 |
| 9 | | 0.3 | 0.2 | 0.4 | 0.6 | 0.5 | 0.6 | 0.8 | 1.2 | 0.7 | 0.8 | 0.8 | 0.6 | |
| 10 | | 0.3 | 0.1 | 0.3 | 0.5 | 0.6 | 0.6 | 0.6 | 1.0 | 0.7 | 0.7 | 0.8 | 0.6 | |
| 11 | | 0.2 | 0.1 | 0.2 | 0.5 | 0.8 | 0.5 | 0.5 | 1.0 | 0.8 | 0.7 | 2.8 | 0.6 | |
| 12 | | 0.3 | 0.1 | 0.2 | 0.3 | 0.8 | 0.5 | 0.5 | 0.9 | 0.7 | 0.6 | 1.3 | 0.6 | |
| 13 | 0.6 | 0.4 | 0.1 | 0.2 | 0.3 | 2.8 | 0.5 | 0.5 | 0.8 | 0.7 | 0.6 | 1.0 | 0.8 | |
| 14 | 0.4 | 0.3 | 0.2 | 0.2 | 0.3 | 1.7 | 0.5 | 0.5 | 1.0 | 0.7 | 0.6 | 0.8 | 0.8 | Ended September 8th, 1892. |
| 15 | 0.4 | 0.2 | 0.2 | 0.2 | 0.5 | 1.1 | 0.7 | 0.5 | 0.8 | 0.7 | 0.5 | 0.8 | 0.6 | |
| 16 | 0.4 | 0.2 | 0.1 | 0.2 | 0.7 | 0.8 | 0.6 | 0.5 | 0.8 | 0.7 | 0.6 | 0.8 | 0.6 | |
| 17 | 0.4 | 0.2 | 0.1 | 0.2 | 0.5 | 0.8 | 0.5 | 0.5 | 0.8 | 1.7 | 0.6 | 0.8 | 0.6 | |
| 18 | 0.4 | 0.2 | 0.1 | 0.2 | 0.4 | 1.1 | 0.5 | 0.5 | 0.8 | 1.7 | 0.6 | 0.8 | 0.6 | |
| 19 | 0.4 | 0.2 | 0.2 | 0.2 | 0.4 | 1.5 | 0.5 | 0.5 | 0.8 | 0.7 | 0.6 | 0.8 | 0.5 | |
| 20 | 0.6 | 0.2 | 0.3 | 0.2 | 0.4 | 1.1 | 0.6 | 0.5 | 0.8 | 0.7 | 0.8 | 0.8 | 0.5 | |
| 21 | 0.5 | 0.2 | 0.1 | 0.2 | 0.3 | 0.8 | 0.8 | 0.5 | 0.8 | 0.7 | 1.5 | 0.8 | 0.5 | |
| 22 | 0.3 | 0.2 | 0.2 | 0.3 | 0.3 | 0.8 | 0.7 | 0.5 | 0.8 | 0.7 | 1.5 | 0.8 | 0.4 | |
| 23 | 0.3 | 0.2 | 0.2 | 0.4 | 0.4 | 0.8 | 0.6 | 0.6 | 0.8 | 0.7 | 1.0 | 0.8 | 0.5 | |
| 24 | 0.6 | 0.2 | 0.2 | 0.6 | 0.4 | 0.8 | 0.6 | 0.8 | 0.8 | 0.7 | 0.8 | 0.8 | 0.5 | |
| 25 | 0.5 | 0.2 | 0.2 | 0.5 | 0.4 | 0.8 | 0.6 | 1.2 | 0.8 | 0.7 | 0.8 | 0.8 | 0.5 | |
| 26 | 0.4 | 0.2 | 0.2 | 0.4 | 0.4 | 0.7 | 0.5 | 0.8 | 0.8 | 0.6 | 0.8 | 0.8 | 0.5 | |
| 27 | 0.6 | 0.2 | 0.2 | 0.4 | 0.4 | 0.7 | 0.5 | 0.7 | 0.8 | 0.6 | 0.8 | 0.8 | 0.5 | |
| 28 | 0.7 | 0.2 | 0.2 | 0.3 | 0.4 | 0.6 | 0.5 | 0.7 | 0.8 | 0.6 | 0.8 | 0.7 | 0.8 | |
| 29 | 0.4 | 0.3 | 0.2 | 0.3 | 0.4 | 0.6 | 0.5 | 0.6 | 0.8 | 0.6 | 0.8 | 0.7 | 0.7 | |
| 30 | 0.4 | 0.2 | 0.2 | 0.3 | 0.4 | 0.5 | . . | 0.6 | 0.8 | 0.6 | 0.8 | 0.7 | 0.6 | |
| 31 | 0.3 | . . | 0.2 | . . | 0.3 | 0.5 | . . | 0.5 | . . | 0.6 | . . | 0.7 | 0.5 | |

NOTE — In the original monthly table, the stages given were the heights of the surface above the bottom of the gauge-rod, which point was 2.2 feet below the 0.0 of this table.

## FIG. 2

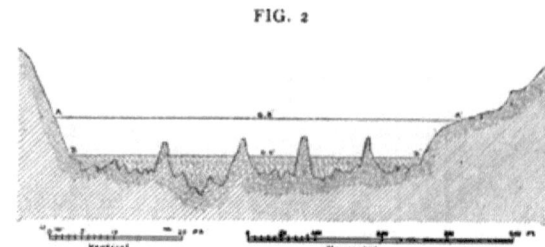

CROSS-SECTION OF THE CHATTAHOOCHEE RIVER, AT ROSWELL BRIDGE, BETWEEN
FULTON AND COBB COUNTIES, GEORGIA

## FLOW OF THE CHATTAHOOCHEE RIVER, AT ROSWELL BRIDGE

| No. | Date of Measurement | Stage | Area in Sq. Ft. of Cross-Section | Velocity in Feet per Second | Discharge in Cu. Ft. per Sec. | REMARKS |
|---|---|---|---|---|---|---|
| 1 | April 22, 1891 | 2.0 | 1,960.4 | 3.70 | 7,253.5 | Measured by C. C. Anderson. |
| 2 | April 12, 1892 | 1.2 | 1,913.0 | 2.85 | 5,452.0 | Measured by C. C. Anderson. |
| 3 | July 2, 1892 | 0.4 | 987.2 | 3.22 | 3,178.7 | Measured by C. C. Anderson. |
| 4 | . . . . . . | 0.0 | 770.4 | 2.84 | 2,190.5 | Calculated. |

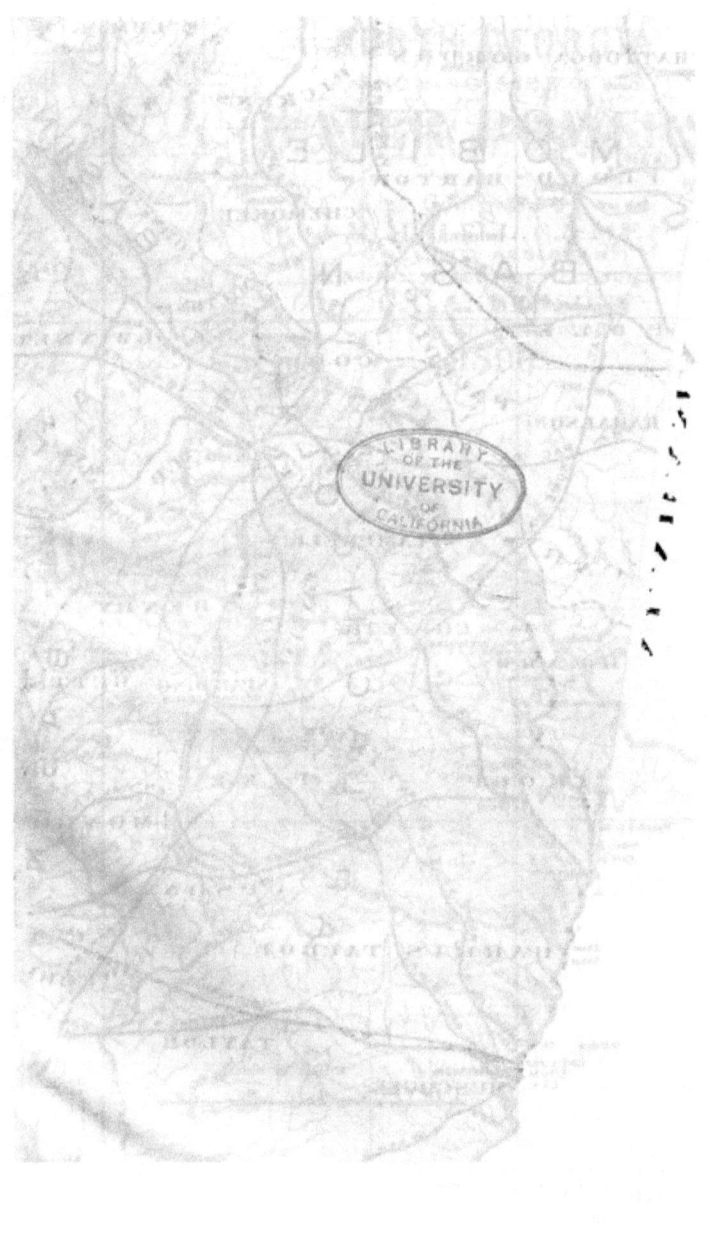

# GEOLOGICAL SURVEY OF GEORGIA

W. S. YEATES, State Geologist

## HYDROGRAPHIC MAP
### OF
## NORTH GEORGIA

SHOWING SOME OF ITS

## WATER POWERS

Compiled by
B. M. HALL, C. & M. E.
SPECIAL ASSISTANT
PRINCIPALLY FROM THE FIELD NOTES OF
G. C. ANDERSON
LATE ASSISTANT GEOLOGIST
1896

## TABLE II
### DAILY FLUCTUATIONS IN FEET AND TENTHS
*Lowest Observed Stage = 0.0*
### THE CHATTAHOOCHEE RIVER AT ROSWELL BRIDGE, COBB AND FULTON COUNTIES, GEORGIA

| Date | 1891 | | | | | 1892 | | | | | | | | |
|---|---|---|---|---|---|---|---|---|---|---|---|---|---|---|
| | Aug. | Sept. | Oct. | Nov. | Dec. | Jan. | Feb. | Mar. | Apr. | May | June | July | Aug. | Sept. |
| 1 | .. | .. | | 0.2 | | 0.4 | 0.5 | 0.5 | 1.2 | 0.4 | 0.1 | 0.4 | 0.2 | 0.1 |
| 2 | .. | .. | | 0.2 | | 0.3 | 0.5 | 0.4 | 1.1 | 0.4 | 0.1 | 0.4 | 0.6 | 0.1 |
| 3 | .. | .. | | 0.2 | | 0.4 | 0.5 | 0.3 | 0.9 | 0.4 | 0.9 | 0.4 | 0.4 | 0.0 |
| 4 | .. | .. | Begun October 10th, 1891. | 0.2 | No Record. | 1.2 | 0.5 | 0.3 | 0.8 | 0.3 | 2.0 | 0.4 | 0.2 | 0.0 |
| 5 | .. | .. | | 0.3 | | 1.5 | 0.5 | 0.3 | 0.7 | 0.2 | 1.6 | 0.6 | 0.2 | 0.0 |
| 6 | .. | .. | | 0.3 | 1.1 | 2.2 | 0.5 | 0.3 | 2.2 | 0.2 | 1.5 | 1.9 | 0.1 | 0.0 |
| 7 | .. | .. | | 0.2 | 1.2 | 2.0 | 0.5 | 0.3 | 5.0 | 0.2 | 0.8 | 1.0 | 0.1 | 0.0 |
| 8 | .. | .. | | 0.2 | 0.8 | 1.8 | 0.8 | 1.2 | 5.6 | 0.2 | 0.8 | 0.8 | 0.1 | 0.0 |
| 9 | .. | .. | | 0.2 | 0.5 | 1.5 | 0.9 | 2.0 | 4.2 | 0.2 | 1.0 | 0.7 | 0.1 | |
| 10 | .. | .. | 0.2 | 0.3 | 0.4 | 1.2 | 0.8 | 1.2 | 1.7 | 0.5 | 0.6 | 1.1 | 0.2 | |
| 11 | .. | .. | 0.1 | 0.4 | 0.3 | 1.1 | 0.6 | 1.1 | 1.2 | 0.8 | 0.3 | 2.8 | 0.2 | |
| 12 | .. | .. | 0.2 | 0.6 | 0.3 | 2.1 | 0.5 | 0.9 | 1.2 | 0.9 | 0.2 | 3.2 | 0.2 | |
| 13 | .. | .. | 0.3 | 0.5 | 0.4 | 3.2 | 0.5 | 0.6 | 1.0 | 0.6 | 0.2 | 2.2 | 0.1 | |
| 14 | .. | .. | 0.4 | 0.5 | 0.3 | 5.4 | 0.5 | 0.5 | 0.9 | 0.4 | 0.2 | 1.2 | 0.0 | |
| 15 | .. | .. | 1.5 | 0.3 | 0.2 | 6.5 | 0.7 | 0.4 | 1.1 | 0.2 | 0.2 | 1.0 | 0.0 | |
| 16 | .. | .. | 1.5 | 0.3 | 0.8 | 6.8 | 0.7 | 0.4 | 1.1 | 0.2 | 0.1 | 0.8 | 0.0 | |
| 17 | .. | .. | 1.3 | 0.3 | 1.2 | | 0.6 | 0.4 | 0.7 | 0.2 | 0.1 | 0.7 | 0.1 | Ended September 8th, 1892. |
| 18 | .. | .. | 1.2 | 0.3 | 0.5 | | 0.6 | 0.4 | 0.6 | 0.3 | 0.1 | 0.6 | 0.4 | |
| 19 | .. | .. | 1.1 | 0.3 | 0.5 | | 0.5 | 0.6 | 0.6 | 0.8 | 0.2 | 0.6 | 0.6 | |
| 20 | .. | .. | 1.2 | 0.3 | | | 0.6 | 0.6 | 1.2 | 0.6 | 0.2 | 0.9 | 0.4 | |
| 21 | .. | .. | 1.1 | 0.3 | | | 1.8 | 0.5 | 0.9 | 0.5 | 1.1 | 0.7 | 0.2 | |
| 22 | .. | .. | 0.7 | 0.4 | | | 2.3 | 0.5 | 0.9 | 0.4 | 1.2 | 0.7 | 0.2 | |
| 23 | .. | .. | 0.3 | 0.8 | No Record. | | 1.4 | 0.5 | 0.9 | 0.3 | 0.8 | 0.4 | 1.6 | |
| 24 | .. | .. | 0.1 | 1.2 | | No Record. | 1.0 | 0.6 | 0.9 | 0.2 | 2.9 | 0.4 | 1.2 | |
| 25 | .. | .. | 0.0 | 1.2 | | | 0.7 | 0.9 | 0.7 | 0.2 | 1.2 | 1.1 | 1.2 | |
| 26 | .. | .. | 0.2 | 1.2 | | | 0.6 | 2.6 | 0.7 | 0.2 | 1.0 | 0.5 | 1.0 | |
| 27 | .. | .. | 0.2 | 1.2 | 0.4 | | 0.5 | 3.2 | 0.5 | 0.2 | 1.1 | 0.3 | 0.9 | |
| 28 | .. | .. | 0.1 | 1.2 | 0.3 | | 0.5 | 1.5 | 0.5 | 0.2 | 1.0 | 0.3 | 1.1 | |
| 29 | .. | .. | 0.1 | .. | 0.3 | | 0.5 | 1.4 | 0.5 | 0.2 | 1.2 | 0.2 | 1.3 | |
| 30 | .. | .. | 0.1 | .. | 0.2 | | .. | 1.2 | 0.5 | 0.2 | 0.7 | 0.2 | 0.2 | |
| 31 | .. | .. | 0.2 | .. | 0.3 | | .. | 1.2 | .. | 0.2 | .. | 0.2 | 0.2 | |

NOTE — In the original monthly table, the stages given were the heights of the surface above the bottom of the gauge-rod, which point was 1.8 feet below the 0.0 of this table.

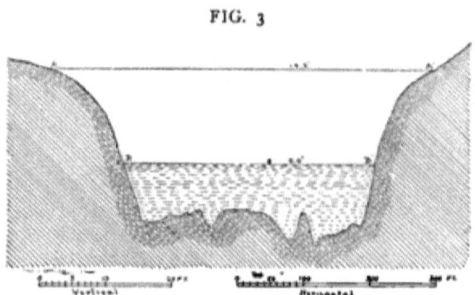

FIG. 3

CROSS-SECTION OF THE CHATTAHOOCHEE RIVER, AT WEST POINT, TROUP COUNTY, GEORGIA

## FLOW OF THE CHATTAHOOCHEE RIVER, AT WEST POINT GEORGIA

| No. | Date of Measurement | Stage | Area in Sq. Ft. of Cross-Section | Velocity in Feet per Second | Discharge in Cu. Ft. per Sec. | REMARKS |
|---|---|---|---|---|---|---|
| 1 | Sept. 26, 1891 | 0.2 | 3,519.5 | 1.54 | 5,414.2 | Measured by C. C. Anderson. |
| 2 | Nov. 24, 1891 | 2.4 | 4,596.0 | 2.00 | 9,192.0 | Measured by C. C. Anderson. |
| 4 | . . . . . . | 0.0 | 3,400.0 | 1.45 | 4,939.5 | Calculated. |

## TABLE III
### DAILY FLUCTUATIONS IN FEET AND TENTHS
*Lowest Observed Stage = 0.0*
#### THE CHATTAHOOCHEE RIVER AT WEST POINT, GEORGIA

| Date | 1891 | | | | 1892 | | | | | | | | |
|---|---|---|---|---|---|---|---|---|---|---|---|---|---|
| | Sept. | Oct. | Nov. | Dec. | Jan. | Feb. | Mar. | April | May | June | July | Aug. | Sept. |
| 1 | | 0.5 | 0.3 | 1.0 | | 1.6 | 1.5 | 2.1 | 1.6 | 0.1 | 1.7 | 0.3 | 0.6 |
| 2 | | 0.6 | 0.3 | 1.3 | | 1.4 | 1.4 | 1.2 | 1.9 | 0.9 | 1.3 | 0.3 | 0.3 |
| 3 | | 0.6 | 0.3 | 1.3 | | 1.1 | 1.3 | 1.3 | 0.9 | 1.1 | 1.0 | 0.4 | 0.2 |
| 4 | | 0.6 | 0.3 | 1.4 | | 1.3 | 1.7 | 1.3 | 0.9 | 3.6 | 0.9 | 0.5 | 0.2 |
| 5 | | 0.6 | 0.4 | 2.1 | | 1.3 | 1.8 | 1.1 | 0.9 | 4.6 | 0.8 | 0.5 | 0.3 |
| 6 | | 0.4 | 0.4 | 2.2 | | 1.3 | 1.1 | 0.9 | 0.9 | 3.6 | 1.0 | 0.5 | 0.0 |
| 7 | | 0.4 | 0.4 | 2.4 | | 2.3 | 1.3 | 7.6 | 0.8 | 2.9 | 1.8 | 0.5 | 0.0 |
| 8 | Begun Sept. 25th, 1891. | 0.5 | 0.4 | 3.3 | No Record. | 2.3 | 6.3 | 12.8 | 0.8 | 1.1 | 1.3 | 0.9 | 0.0 |
| 9 | | 0.4 | 0.4 | 2.6 | | 4.7 | 6.3 | 13.4 | 0.8 | 1.3 | 1.0 | 1.0 | |
| 10 | | 0.6 | 0.5 | 3.0 | | 4.7 | 5.7 | 14.2 | 1.1 | 1.1 | 3.3 | 1.3 | |
| 11 | | 0.6 | 1.5 | 2.7 | | 4.0 | 3.7 | 8.9 | 1.4 | 1.4 | 3.2 | 0.9 | |
| 12 | | 0.5 | 1.5 | 1.6 | | 3.7 | 3.1 | 3.3 | 1.1 | 1.5 | 5.3 | 0.6 | |
| 13 | | 0.4 | 1.2 | 1.4 | | 2.7 | 3.3 | 2.3 | 0.9 | 1.4 | 9.3 | 0.4 | |
| 14 | | 0.4 | 1.2 | 1.3 | | 1.3 | 1.6 | 2.2 | 0.8 | 1.1 | 8.3 | 0.3 | |
| 15 | | 0.4 | 1.8 | 3.0 | | 1.3 | 1.5 | 2.3 | 0.8 | 1.0 | 2.3 | 0.0 | |
| 16 | | 0.5 | 1.6 | 2.2 | | 1.7 | 1.4 | 1.9 | 1.1 | 0.9 | 1.0 | 0.0 | Ended Sept. 8th, 1892. |
| 17 | | 0.3 | 1.6 | 3.7 | | 1.6 | 1.5 | 2.3 | 1.2 | 0.9 | 0.9 | 0.3 | |
| 18 | | 0.4 | 1.6 | 3.3 | | 1.6 | 2.3 | 2.9 | 1.2 | 0.9 | 1.3 | 2.5 | |
| 19 | | 0.3 | 1.6 | 2.1 | | 1.4 | 2.3 | 2.0 | 1.6 | 0.9 | 1.9 | 3.2 | |
| 20 | | 0.3 | 1.6 | 2.1 | | 1.6 | 2.5 | 1.4 | 1.4 | 1.1 | 2.3 | 2.0 | |
| 21 | | 0.3 | 1.6 | 2.1 | | 3.3 | 2.4 | 2.0 | 1.1 | 1.9 | 2.4 | 1.0 | |
| 22 | | 0.3 | 1.6 | 2.6 | | 5.2 | 1.3 | 2.5 | 1.1 | 3.0 | 1.2 | 2.0 | |
| 23 | | 0.3 | 1.9 | 1.6 | | 5.2 | 1.1 | 2.1 | 1.1 | 4.9 | 1.1 | 1.9 | |
| 24 | | 0.3 | 2.4 | 1.4 | | 3.0 | 2.1 | 2.4 | 0.8 | 4.3 | 0.3 | 2.5 | |
| 25 | 0.5 | 0.3 | 3.3 | 1.6 | | 2.5 | 5.8 | 2.0 | 0.8 | 2.1 | 0.4 | 2.0 | |
| 26 | 0.5 | 0.3 | 2.1 | 2.6 | | 2.0 | 10.0 | 2.4 | 0.8 | 1.1 | 0.5 | 1.6 | |
| 27 | 0.5 | 0.3 | 1.7 | 1.2 | 3.6 | 1.9 | 12.8 | 2.0 | 0.8 | 2.1 | 0.5 | 1.0 | |
| 28 | 0.5 | 0.3 | 1.3 | 1.3 | 3.3 | 1.6 | 10.0 | 1.6 | 0.9 | 2.5 | 0.8 | 1.7 | |
| 29 | 0.5 | 0.3 | 1.3 | 1.2 | 3.1 | 1.5 | 6.4 | 1.6 | 0.7 | 2.1 | 0.3 | 1.2 | |
| 30 | 0.5 | 0.3 | 1.3 | 1.2 | 2.3 | . . | 3.2 | 1.5 | 1.1 | 2.0 | 0.3 | 1.0 | |
| 31 | . . | 0.3 | . . | 1.5 | 2.1 | . . | 2.2 | . . | 1.1 | . . | 0.4 | 1.0 | |

NOTE — In the original monthly table, the stages given were the heights of the surface above the bottom of the gauge-rod, which point was the same level as the 0.0 of this table.

FIG. 4

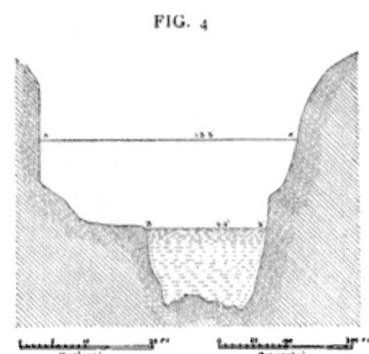

CROSS-SECTION OF THE CHATTAHOOCHEE RIVER, AT COLUMBUS, GEORGIA

## FLOW OF THE CHATTAHOOCHEE RIVER, AT COLUMBUS, GEORGIA

| No. | Date of Measurement | Stage | Area in Sq. Ft. of Cross-Section | Velocity in Feet per Second | Discharge in Cu. Ft. per Sec. | REMARKS |
|---|---|---|---|---|---|---|
| 1 | Aug. 24, 1891 | 1.5 | 2,365.75 | 2.68 | 6,348.1 | Measured by C. C. Anderson. |
| 2 | Jan. 14, 1892 | 13.3 | 8,307.50 | Vel. not taken. | . . . . | Maximum. |
| 3 | Nov. 29, 1892 | 4.8 | 4,083.06+ | 4.94 | 20,190.8 | Measured by C. C. Anderson. |
| 4 | Oct. 29, 1891 | 0.0 | . . . . | . . . . | 5,221.1 | Stated in table by C. C. Anderson. |

## TABLE IV
### DAILY FLUCTUATIONS IN FEET AND TENTHS
*Lowest Observed Stage = 0.0*
### THE CHATTAHOOCHEE RIVER AT COLUMBUS, GEORGIA

| Date | 1891 | | | | | 1892 | | | | | | | | |
|---|---|---|---|---|---|---|---|---|---|---|---|---|---|---|
| | Aug. | Sept. | Oct. | Nov. | Dec. | Jan. | Feb. | Mar. | Apr. | May | June | July | Aug. | Sept. |
| 1 | | 0.9 | 0.4 | 0.6 | 0.9 | 1.0 | 1.8 | 1.8 | 2.5 | 2.0 | 1.1 | 1.6 | 0.9 | 0.9 |
| 2 | | 0.8 | 0.4 | 0.2 | 0.6 | 1.6 | 1.8 | 1.8 | 2.4 | 1.6 | 1.3 | 1.5 | 0.8 | 0.9 |
| 3 | | 1.0 | 0.4 | 0.2 | 0.5 | 2.1 | 1.7 | 1.7 | 2.6 | 1.6 | 1.3 | 1.7 | 1.0 | 0.8 |
| 4 | | 1.8 | 0.8 | 0.1 | 0.9 | 1.8 | 1.7 | 1.6 | 2.3 | 1.5 | 2.5 | 1.7 | 1.0 | 1.2 |
| 5 | | 1.3 | 0.3 | 0.1 | 1.5 | 1.8 | 1.6 | 1.5 | 2.2 | 1.5 | 3.0 | 1.1 | 0.9 | 0.5 |
| 6 | | 1.5 | 0.3 | 0.1 | 1.8 | 1.6 | 1.6 | 1.8 | 2.0 | 1.4 | 2.7 | 1.8 | 0.8 | 0.4 |
| 7 | | 1.1 | 0.3 | 0.2 | 1.7 | 2.0 | 1.9 | 1.6 | 2.4 | 1.6 | 2.2 | 1.3 | 1.4 | 0.4 |
| 8 | Begun August 24th, 1891. | 0.9 | 0.2 | 0.6 | 2.0 | 1.7 | 2.3 | 3.8 | 7.3 | 1.8 | 2.0 | 2.7 | 1.2 | 0.4 |
| 9 | | 0.6 | 0.3 | 0.1 | 1.9 | 1.7 | 4.8 | 4.8 | 7.8 | 1.3 | 1.9 | 2.0 | 1.3 | |
| 10 | | 0.4 | 0.3 | 0.1 | 1.8 | 1.9 | 4.0 | 3.9 | 7.9 | 1.3 | 1.8 | 4.0 | 1.2 | |
| 11 | | 0.4 | 0.9 | 0.7 | 1.5 | 1.7 | 2.9 | 3.2 | 7.5 | 1.3 | 1.7 | 4.2 | 0.9 | |
| 12 | | 0.5 | 0.4 | 0.9 | 1.1 | 3.3 | 2.2 | 2.4 | 4.2 | 1.2 | 1.5 | 4.3 | 1.2 | |
| 13 | | 1.5 | 0.4 | 0.6 | 1.3 | 5.4 | 1.9 | 2.3 | 3.9 | 1.5 | 1.2 | 5.0 | 1.0 | |
| 14 | | 1.1 | 0.4 | 0.8 | 1.0 | 13.3 | 2.1 | 1.9 | 3.6 | 1.5 | 1.0 | 4.9 | 1.3 | |
| 15 | | 1.0 | 0.4 | 1.1 | 0.8 | 12.4 | 2.0 | 1.8 | 3.5 | 1.8 | 0.9 | 3.4 | 0.8 | |
| 16 | | 1.0 | 0.4 | 0.4 | 1.6 | 9.7 | 2.1 | 1.7 | 3.4 | 1.2 | 0.8 | 2.6 | 0.9 | Ended September 8th, 1892. |
| 17 | | 0.9 | 0.3 | 0.4 | 1.7 | 8.7 | 1.9 | 1.6 | 3.5 | 1.2 | 0.7 | 2.1 | 0.8 | |
| 18 | | 0.6 | 0.6 | 0.6 | 1.8 | 8.4 | 1.9 | 1.9 | 3.2 | 1.3 | 0.8 | 2.1 | 1.4 | |
| 19 | | 0.5 | 0.0 | 0.6 | 1.9 | 4.6 | 1.8 | 1.6 | 3.1 | 1.4 | 1.4 | 2.0 | 1.3 | |
| 20 | | 0.9 | 0.0 | 0.6 | 2.0 | 7.9 | 1.8 | 1.9 | 3.0 | 1.7 | 0.9 | 2.8 | 3.0 | |
| 21 | | 0.5 | 0.1 | 0.6 | 1.5 | 7.8 | 3.8 | 1.6 | 3.0 | 1.5 | 0.9 | 2.7 | 2.5 | |
| 22 | | 0.4 | 0.1 | 1.0 | 1.3 | 6.4 | 3.8 | 1.6 | 2.6 | 1.8 | 2.2 | 2.4 | 1.7 | |
| 23 | | 0.3 | 0.0 | 1.7 | 1.2 | 4.3 | 4.0 | 1.6 | 2.2 | 1.3 | 1.9 | 1.7 | 2.5 | |
| 24 | 1.5 | 0.2 | 0.0 | 1.6 | 1.1 | 3.2 | 3.1 | 3.4 | 2.3 | 1.4 | 2.1 | 2.1 | 2.7 | |
| 25 | 1.5 | 0.2 | 0.3 | 1.7 | 1.5 | 2.6 | 2.4 | 4.6 | 2.0 | 1.2 | 2.7 | 1.7 | 2.6 | |
| 26 | 1.3 | 0.1 | 0.0 | 2.0 | 1.5 | 2.4 | 2.1 | 11.2 | 1.9 | 1.2 | 2.2 | 1.7 | 2.1 | |
| 27 | 1.2 | 0.7 | 0.0 | 1.2 | 1.6 | 2.2 | 2.0 | 9.9 | 1.8 | 1.1 | 1.5 | 1.9 | 2.1 | |
| 28 | 1.3 | 0.4 | 0.0 | 0.9 | 1.3 | 2.1 | 2.1 | 7.1 | 1.7 | 1.1 | 3.0 | 1.2 | 2.2 | |
| 29 | 1.1 | 0.4 | 0.0 | 1.4 | 1.3 | 2.0 | 1.8 | 4.8 | 1.7 | 1.4 | 2.5 | 1.2 | 1.4 | |
| 30 | 1.3 | 0.4 | 0.0 | 0.9 | 1.2 | 1.9 | .. | 3.4 | 1.7 | 1.2 | 1.9 | 1.0 | 1.3 | |
| 31 | 0.9 | .. | 0.1 | .. | 1.1 | 2.1 | .. | 2.8 | .. | 1.2 | .. | 0.8 | 1.0 | |

NOTE.—In the original monthly table, the stages given were the heights of the surface above the bottom of the gauge-rod, which point was 1.6 feet below the 0.0 of this table.

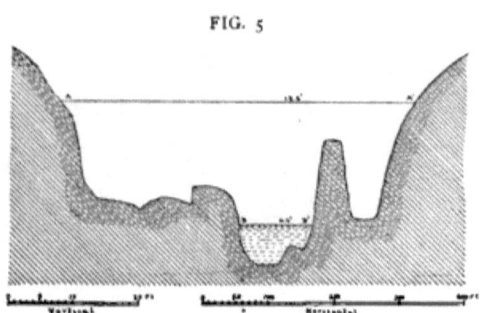

FIG. 5

CROSS-SECTION OF THE FLINT RIVER, AT SULLIVAN'S MILL, PIKE COUNTY, GEORGIA

## FLOW OF THE FLINT RIVER, AT SULLIVAN'S MILL, PIKE COUNTY, GEORGIA

| No. | Date of Measurement | Stage | Area in Sq. Ft. of Cross-Section | Velocity in Feet per Second | Discharge in Cu. Ft. per Sec. | REMARKS |
|---|---|---|---|---|---|---|
| 1 | Mar. 16, 1891 | 2.6 | 674.5 | 0.91 | 612.6 | Measured by C. C. Anderson. |
| 2 | . . . . . . | 0.0 | 375.0 | 0.66 | 250.0 | Calculated. |

## TABLE V
### DAILY FLUCTUATIONS IN FEET AND TENTHS
*Lowest Observed Stage = 0.0*

THE FLINT RIVER, SULLIVAN'S MILL, PIKE COUNTY, GEORGIA
NEAR ERIN P. O., MERIWETHER COUNTY

| Date | 1891 | | | | | | | | 1892 | | | | | | | | |
|---|---|---|---|---|---|---|---|---|---|---|---|---|---|---|---|---|---|
| | May | June | July | Aug. | Sept. | Oct. | Nov. | Dec. | Jan. | Feb. | Mar. | Apr. | May | June | July | Aug. | Sept. |
| 1 | | 2.5 | 3.1 | 2.5 | 2.4 | 0.6 | 0.7 | 1.7 | 1.4 | 2.2 | 2.2 | 4.0 | 2.0 | 1.3 | 2.3 | 1.4 | 2.4 |
| 2 | | 2.4 | 2.7 | 2.4 | 2.4 | 0.3 | 1.0 | 1.5 | 1.5 | 2.3 | 2.2 | 3.1 | 1.9 | 1.2 | 3.2 | 1.3 | 1.8 |
| 3 | | 2.3 | 3.1 | 2.4 | 2.3 | 0.4 | 1.0 | 1.4 | 2.2 | 2.3 | 2.2 | 2.8 | 1.9 | 1.4 | 2.5 | 1.7 | 1.8 |
| 4 | | 2.2 | 2.8 | 2.5 | 2.5 | 0.2 | 1.3 | 2.2 | 1.8 | 1.8 | 2.1 | 2.5 | 1.9 | 2.1 | 2.1 | 1.8 | 1.5 |
| 5 | | 2.2 | 2.4 | 2.6 | 2.6 | 0.0 | 1.3 | 3.0 | 1.4 | 1.7 | 2.0 | 2.3 | 1.8 | 2.4 | 2.4 | 2.0 | 1.6 |
| 6 | | 2.2 | 2.4 | 3.1 | 2.7 | 0.4 | 1.3 | 2.5 | 2.2 | 1.5 | 1.9 | 2.2 | 1.8 | 2.6 | 2.6 | 1.8 | 1.5 |
| 7 | | 2.1 | 2.5 | 3.3 | 3.1 | 0.2 | 1.3 | 2.5 | 3.1 | 1.5 | 2.1 | 3.8 | 1.8 | 2.2 | 2.8 | 1.5 | 1.4 |
| 8 | | 2.5 | 2.3 | 3.1 | 2.7 | 0.0 | 1.0 | 2.6 | 2.9 | 4.8 | 2.9 | 10.1 | 1.8 | 2.0 | 2.4 | 1.2 | 1.3 |
| 9 | Begun May 15th, 1891. | 4.9 | 2.2 | 3.3 | 2.5 | 0.6 | 1.0 | 2.0 | 1.5 | 7.1 | 4.2 | 13.5 | 1.7 | 1.9 | 2.2 | 1.6 | |
| 10 | | 3.2 | 1.9 | 2.9 | 2.2 | 0.7 | 0.8 | 1.6 | 1.2 | 8.5 | 4.5 | 10.5 | 1.7 | 1.8 | 3.4 | 2.4 | |
| 11 | | 7.2 | 1.7 | 2.7 | 2.1 | 0.6 | 2.1 | 1.5 | 3.0 | 7.1 | 4.2 | 6.3 | 1.8 | 2.4 | 4.6 | 2.3 | |
| 12 | | 4.7 | 1.5 | 2.4 | 2.3 | 0.6 | 1.6 | 1.5 | 4.2 | 5.2 | 3.2 | 4.3 | 1.8 | 2.1 | 3.7 | 2.2 | |
| 13 | | 5.1 | 1.4 | 2.4 | 2.0 | 0.7 | 1.5 | 1.4 | 7.2 | 3.4 | 3.1 | 3.3 | 1.8 | 1.8 | 3.5 | 2.0 | |
| 14 | | 5.3 | 1.2 | 2.3 | 1.8 | 0.7 | 1.4 | 1.4 | 14.2 | 2.9 | 2.8 | 2.8 | 1.8 | 1.6 | 4.2 | 1.8 | |
| 15 | 2.6 | 4.7 | 1.2 | 2.5 | 1.5 | 0.9 | 1.4 | 1.5 | 18.6 | 2.8 | 2.3 | 2.6 | 1.7 | 1.5 | 4.4 | 1.6 | |
| 16 | 2.5 | 4.4 | 1.2 | 2.7 | 1.4 | 0.7 | 1.2 | 1.5 | 14.2 | 3.4 | 2.0 | 2.2 | 1.8 | 1.4 | 3.2 | 1.5 | |
| 17 | 2.4 | 2.8 | 1.1 | 3.0 | 1.6 | 0.6 | 1.2 | 2.0 | 12.4 | 3.0 | 1.8 | 2.1 | 1.9 | 1.3 | 4.4 | 2.8 | |
| 18 | 2.4 | 2.2 | 1.1 | 2.7 | 1.7 | 0.5 | 1.1 | 2.4 | 7.2 | 3.1 | 2.4 | 1.9 | 2.1 | 1.2 | 3.4 | 8.5 | |
| 19 | 2.5 | 2.4 | 1.1 | 2.5 | 1.9 | 0.4 | 1.1 | 2.2 | 7.8 | 2.8 | 3.0 | 1.9 | 2.4 | 1.2 | 2.9 | 7.8 | |
| 20 | 2.9 | 2.4 | 1.3 | 2.6 | 1.7 | 0.3 | 1.1 | 2.4 | 12.4 | 2.3 | 2.6 | 1.8 | 2.5 | 1.7 | 3.2 | 5.7 | Ended Sept. 8th, 1892. |
| 21 | 3.0 | 2.4 | 1.4 | 2.7 | 1.5 | 0.3 | 1.1 | 2.2 | 15.2 | 3.2 | 2.4 | 1.8 | 2.3 | 2.8 | 3.2 | 5.2 | |
| 22 | 2.7 | 3.1 | 1.6 | 3.1 | 1.3 | 0.3 | 1.1 | 2.0 | 13.5 | 4.6 | 2.1 | 2.0 | 2.1 | 2.3 | 3.6 | 3.2 | |
| 23 | 2.6 | 3.4 | 1.8 | 3.5 | 1.0 | 0.2 | 2.2 | 1.9 | 10.4 | 6.8 | 1.8 | 1.9 | 2.0 | 2.1 | 3.4 | 6.6 | |
| 24 | 2.4 | 4.0 | 2.2 | 7.1 | 0.9 | 0.2 | 2.5 | 1.8 | 7.4 | 5.8 | 2.7 | 1.8 | 1.9 | 2.0 | 2.8 | 6.8 | |
| 25 | 2.7 | 4.7 | 2.4 | 7.8 | 0.8 | 0.2 | 2.4 | 1.7 | 5.2 | 4.5 | 5.5 | 1.7 | 1.8 | 2.2 | 2.5 | 5.8 | |
| 26 | 2.8 | 4.2 | 2.6 | 7.3 | 0.7 | 0.3 | 2.0 | 1.6 | 3.4 | 3.1 | 13.0 | 1.7 | 1.6 | 2.4 | 2.2 | 7.8 | |
| 27 | 2.8 | 3.4 | 2.7 | 5.2 | 0.6 | 0.4 | 1.7 | 2.2 | 3.0 | 3.2 | 14.8 | 1.8 | 1.5 | 2.8 | 1.8 | 5.8 | |
| 28 | 3.1 | 2.7 | 3.1 | 4.7 | 0.5 | 0.5 | 1.5 | 2.3 | 2.8 | 2.8 | 15.0 | 1.8 | 1.5 | 4.5 | 1.8 | 4.8 | |
| 29 | 3.3 | 2.5 | 3.0 | 3.9 | 0.4 | 0.6 | 1.7 | 1.7 | 2.7 | 2.4 | 13.1 | 1.9 | 1.6 | 3.7 | 1.7 | 3.0 | |
| 30 | 3.1 | 2.2 | 2.7 | 3.3 | 0.3 | 0.7 | 1.8 | 1.5 | 2.6 | . . | 7.9 | 2.0 | 1.6 | 3.7 | 1.6 | 3.1 | |
| 31 | 2.6 | . . | 2.5 | 2.7 | . . | 0.7 | . . | 1.4 | 2.6 | . . | 5.2 | . . | 1.5 | . . | 1.5 | 3.2 | |

NOTE — In the original monthly table, the stages given were the heights of the surface above the bottom of the gauge-rod, which point was 1.2 feet below the 0.0 of this table.

FIG. 6

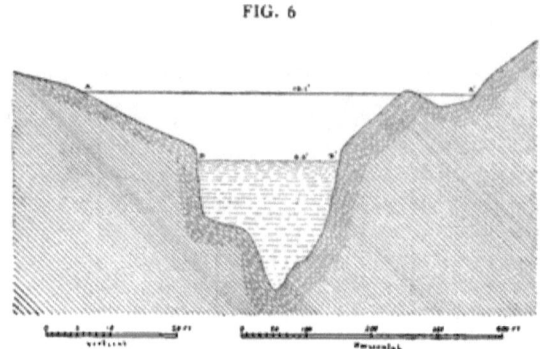

CROSS-SECTION OF THE FLINT RIVER, AT THE MACON & BIRMINGHAM R. R. BRIDGE, MERIWETHER COUNTY, GEORGIA

## FLOW OF FLINT RIVER, AT THE MACON & BIRMINGHAM R. R. BRIDGE, MERIWETHER COUNTY, GEORGIA

| No. | Date of Measurement | Stage | Area in Sq. Ft. of Cross-Section | Velocity in Feet per Second | Discharge in Cu. Ft. per Sec. | REMARKS |
|---|---|---|---|---|---|---|
| 1 | Aug. 25, 1891 | 3.3 | 3,051.8+ | 2.00 | 6,103.6 | Measured by C. C. Anderson. |
| 2 | April 1, 1892 | 2.2 | 1,904.7 | 1.84 | 3,497.9 | Measured by C. C. Anderson. |

Data not sufficient for calculating minimum discharge.

WATER POWERS OF GEORGIA

INDIAN ARROW RAPIDS, THE HEAD OF TALLULAH FALLS, GEORGIA.

## TABLE VI

### DAILY FLUCTUATIONS IN FEET AND TENTHS

*Lowest Observed Stage = 0.0*

### THE FLINT RIVER AT THE MACON AND BIRMINGHAM R. R. BRIDGE, MERIWETHER COUNTY, GEORGIA

| Date | 1891 | | | | 1892 | | | | | | | | |
|---|---|---|---|---|---|---|---|---|---|---|---|---|---|
| | Aug. | Sept. | Oct. | Nov. | Dec. | Jan. | Feb. | Mar. | Apr. | May | June | July | Aug. | Sept. |
| 1 |  | 0.9 | 0.2 | 0.3 | 0.8 | 0.8 | 1.4 | 1.3 | 2.2 | 0.9 | 0.4 | 0.8 | 0.4 | 1.2 |
| 2 |  | 3.4 | 0.2 | 0.2 | 0.7 | 1.3 | 1.4 | 1.3 | 1.9 | 0.8 | 0.4 | 0.7 | 0.3 | 0.8 |
| 3 |  | 2.9 | 0.2 | 0.2 | 0.7 | 1.3 | 1.2 | 1.2 | 1.7 | 0.8 | 1.0 | 0.6 | 0.3 | 0.7 |
| 4 |  | 3.0 | 0.2 | 0.3 | 1.2 | 1.2 | 1.2 | 1.2 | 1.6 | 0.8 | 1.8 | 0.4 | 0.8 | 0.6 |
| 5 |  | 2.1 | 0.2 | 0.3 | 1.1 | 1.0 | 1.1 | 1.1 | 1.5 | 0.7 | 1.1 | 1.1 | 0.7 | 0.5 |
| 6 |  | 1.1 | 0.2 | 0.4 | 1.6 | 1.3 | 1.7 | 1.1 | 1.5 | 0.7 | 1.2 | 2.3 | 0.4 | 0.4 |
| 7 |  | 0.9 | 0.2 | 0.4 | 1.6 | 1.8 | 6.1 | 1.1 | 1.9 | 0.6 | 1.0 | 1.4 | 0.7 | 0.4 |
| 8 |  | 0.8 | 0.2 | 0.3 | 1.8 | 1.7 | 5.3 | 2.1 | 3.4 | 0.6 | 0.8 | 1.2 | 0.6 | 0.4 |
| 9 |  | 0.7 | 0.2 | 0.3 | 1.5 | 1.4 | 6.1 | 2.4 | 5.5 | 0.6 | 0.6 | 0.9 | 1.9 |  |
| 10 | Begun Aug. 25, 1891. | 0.9 | 0.2 | 0.5 | 1.3 | 1.5 | 4.7 | 2.3 | 5.1 | 0.6 | 0.7 | 3.0 | 1.5 |  |
| 11 |  | 0.7 | 0.2 | 0.9 | 1.2 | 1.6 | 3.4 | 2.0 | 3.2 | 0.7 | 0.6 | 3.2 | 1.2 |  |
| 12 |  | 1.2 | 0.0 | 0.9 | 1.0 | 2.6 | 2.8 | 1.6 | 2.1 | 0.8 | 0.6 | 2.5 | 0.8 |  |
| 13 |  | 1.3 | 0.1 | 0.7 | 0.7 | 4.2 | 2.1 | 1.4 | 1.8 | 0.7 | 0.4 | 1.9 | 0.6 |  |
| 14 |  | 1.5 | 0.1 | 0.7 | 0.8 | 8.8 | 1.7 | 1.3 | 1.6 | 0.6 | 0.3 | 1.9 | 0.4 |  |
| 15 |  | 1.3 | 0.3 | 0.6 | 0.8 | 10.1 | 1.8 | 1.2 | 1.5 | 0.6 | 0.2 | 1.8 | 0.3 |  |
| 16 |  | 1.0 | 0.2 | 0.5 | 1.2 | 9.6 | 1.9 | 1.1 | 1.4 | 0.7 | 0.2 | 1.6 | 0.3 |  |
| 17 |  | 0.9 | 0.1 | 0.5 | 1.6 | 7.2 | 1.8 | 1.1 | 1.3 | 0.7 | 0.2 | 1.6 | 1.7 |  |
| 18 |  | 0.3 | 0.1 | 0.5 | 1.5 | 4.3 | 1.6 | 1.5 | 1.2 | 0.7 | 0.2 | 1.9 | 3.1 | Ended Sept. 8, 1892. |
| 19 |  | 0.2 | 0.0 | 0.5 | 1.3 | 9.1 | 1.5 | 1.6 | 1.2 | 1.0 | 0.6 | 2.2 | 4.4 |  |
| 20 |  | 0.2 | 0.0 | 0.5 | 1.5 | 8.8 | 1.4 | 1.5 | 1.2 | 0.9 | 0.7 | 2.2 | 3.4 |  |
| 21 |  | 0.3 | 0.0 | 0.5 | 1.4 | 7.4 | 2.3 | 1.3 | 1.1 | 0.8 | 1.5 | 1.7 | 2.4 |  |
| 22 |  | 0.2 | 0.1 | 0.4 | 1.2 | 4.8 | 2.7 | 1.2 | 1.2 | 0.7 | 1.5 | 1.7 | 2.3 |  |
| 23 |  | 0.2 | 0.2 | 0.6 | 1.1 | 2.8 | 3.4 | 1.1 | 1.1 | 0.7 | 0.9 | 2.0 | 3.9 |  |
| 24 |  | 0.2 | 0.1 | 1.7 | 1.1 | 2.2 | 2.7 | 2.1 | 1.1 | 0.8 | 0.8 | 1.7 | 3.7 |  |
| 25 | 3.3 | 0.0 | 0.1 | 1.1 | 1.1 | 1.9 | 2.8 | 4.6 | 1.1 | 0.7 | 0.9 | 0.8 | 3.8 |  |
| 26 | 2.4 | 0.2 | 0.0 | 1.0 | 1.1 | 1.7 | 1.7 | 10.0 | 1.0 | 0.6 | 0.9 | 0.9 | 3.1 |  |
| 27 | 1.5 | 0.2 | 0.2 | 0.9 | 1.1 | 1.5 | 1.5 | 9.6 | 1.0 | 0.6 | 1.3 | 0.6 | 3.2 |  |
| 28 | 2.6 | 0.2 | 0.2 | 0.7 | 0.3 | 1.4 | 1.4 | 8.5 | 0.9 | 0.5 | 2.2 | 0.5 | 2.2 |  |
| 29 | 1.5 | 0.2 | 0.2 | 0.8 | 1.0 | 1.4 | 1.4 | 7.9 | 0.9 | 0.5 | 1.9 | 0.4 | 2.6 |  |
| 30 | 1.1 | 0.2 | 0.2 | 0.8 | 0.9 | 1.4 | . . | 4.3 | 0.9 | 0.4 | 1.4 | 0.3 | 1.2 |  |
| 31 | 0.9 | . . | 0.1 | . . | 0.9 | 4.5 | . . | 4.6 | . . | 0.5 | . . | 0.3 | 0.3 |  |

NOTE — In the original monthly table, the stages given were the heights of the surface above the bottom of the gauge-rod, which point was 0.9 foot below the 0.0 of this table.

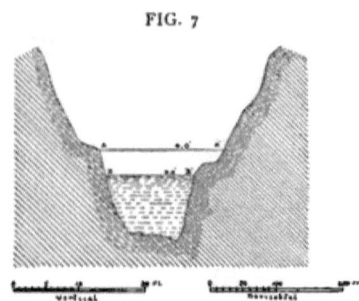

FIG. 7

CROSS-SECTION OF BIG POTATO CREEK, AT NELSON'S MILL, UPSON COUNTY, GEORGIA

## FLOW OF BIG POTATO CREEK, AT NELSON'S MILL, UPSON COUNTY, GEORGIA

| No. | Date of Measurement | Stage | Area in Sq. Ft. of Cross-Section | Velocity in Feet per Second | Discharge in Cu. Ft. per Sec. | REMARKS |
|---|---|---|---|---|---|---|
| 1 | Aug. 18, 1892 | 0.0 | 88.0 | 1.25 | 110.0 | Measured by C. C. Anderson. |

## TABLE VII
### DAILY FLUCTUATIONS IN FEET AND TENTHS
*Lowest Observed Stage = 0.0*
### BIG POTATO CREEK AT NELSON'S MILL, UPSON COUNTY, GEORGIA

| Date | 1891 | | | | 1892 | | | | | | | | |
|---|---|---|---|---|---|---|---|---|---|---|---|---|---|
| | Sept. | Oct. | Nov. | Dec. | Jan. | Feb. | Mar. | April | May | June | July | Aug. | Sept. |
| 1 | 0.0 | 0.0 | 0.0 | 0.0 | 0.1 | 0.2 | 0.2 | 0.5 | 0.1 | 0.0 | 0.1 | 0.1 | 0.1 |
| 2 | 0.0 | 0.0 | 0.0 | 0.0 | 0.2 | 0.2 | 0.2 | 0.5 | 0.1 | 0.0 | 0.1 | 0.1 | 0.1 |
| 3 | 0.0 | 0.0 | 0.0 | 0.0 | 0.2 | 0.2 | 0.2 | 0.4 | 0.1 | 0.1 | 0.1 | 0.1 | 0.1 |
| 4 | 0.0 | 0.0 | 0.0 | 0.1 | 0.2 | 0.2 | 0.2 | 0.4 | 0.1 | 0.1 | 0.1 | 0.2 | 0.1 |
| 5 | 0.0 | 0.0 | 0.0 | 0.1 | 0.1 | 0.2 | 0.1 | 0.3 | 0.1 | 0.1 | 0.1 | 0.1 | 0.1 |
| 6 | 0.0 | 0.0 | 0.0 | 0.2 | 0.2 | 0.2 | 0.1 | 0.3 | 0.1 | 0.1 | 0.1 | 0.1 | 0.1 |
| 7 | 0.0 | 0.0 | 0.0 | 0.2 | 0.2 | 0.2 | 0.2 | 0.3 | 0.1 | 0.1 | 0.1 | 0.2 | 0.1 |
| 8 | 0.0 | 0.0 | 0.0 | 0.2 | 0.2 | 0.6 | 0.6 | 0.5 | 0.1 | 0.1 | 0.2 | 0.2 | 0.1 |
| 9 | 0.0 | 0.0 | 0.0 | 0.3 | 0.2 | 1.1 | 0.5 | 0.5 | 0.1 | 0.3 | 0.5 | 0.4 | |
| 10 | 0.0 | 0.0 | 0.0 | 0.2 | 0.2 | 1.2 | 0.6 | 0.4 | 0.1 | 0.1 | 0.3 | 0.4 | |
| 11 | 0.0 | 0.0 | 0.0 | 0.1 | 0.2 | 0.7 | 0.4 | 0.3 | 0.1 | 0.2 | 1.0 | 0.4 | |
| 12 | 0.0 | 0.0 | 0.0 | 0.1 | 0.5 | 0.6 | 0.3 | 0.3 | 0.1 | 0.1 | 1.1 | 0.2 | |
| 13 | 0.1 | 0.0 | 0.0 | 0.1 | 0.8 | 0.7 | 0.2 | 0.3 | 0.1 | 0.0 | 0.7 | 0.2 | |
| 14 | 0.1 | 0.0 | 0.0 | 0.1 | 1.3 | 0.5 | 0.2 | 0.2 | 0.1 | 0.0 | 0.4 | 0.2 | |
| 15 | 0.1 | 0.0 | 0.0 | 0.0 | 1.4 | 0.6 | 0.2 | 0.2 | 0.1 | 0.0 | 0.3 | 0.2 | |
| 16 | 0.0 | 0.0 | 0.0 | 0.1 | 1.0 | 0.4 | 0.2 | 0.2 | 0.1 | 0.0 | 0.3 | 0.1 | Ended September 8th, 1892. |
| 17 | 0.0 | 0.0 | 0.0 | 0.1 | 0.6 | 0.5 | 0.2 | 0.2 | 0.1 | 0.0 | 0.2 | 0.4 | |
| 18 | 0.0 | 0.0 | 0.0 | 0.1 | 1.1 | 0.4 | 0.3 | 0.2 | 0.1 | 0.0 | 0.2 | 1.3 | |
| 19 | 0.0 | 0.0 | 0.0 | 0.1 | 1.4 | 0.3 | 0.3 | 0.2 | 0.1 | 0.0 | 0.3 | 1.2 | |
| 20 | 0.0 | 0.0 | 0.0 | 0.1 | 1.7 | 0.3 | 0.3 | 0.2 | 0.1 | 0.1 | 0.3 | 0.7 | |
| 21 | 0.0 | 0.0 | 0.0 | 0.1 | 1.6 | 0.9 | 0.2 | 0.1 | 0.1 | 0.1 | 0.7 | 0.5 | |
| 22 | 0.0 | 0.0 | 0.0 | 0.2 | 0.9 | 0.8 | 0.2 | 0.1 | 0.1 | 0.1 | 0.7 | 0.3 | |
| 23 | 0.0 | 0.0 | 0.6 | 0.2 | 0.6 | 0.6 | 0.2 | 0.1 | 0.1 | 0.1 | 0.3 | 0.8 | |
| 24 | 0.0 | 0.0 | 0.2 | 0.1 | 0.6 | 0.5 | 1.0 | 0.1 | 0.1 | 0.1 | 0.3 | 0.8 | |
| 25 | 0.0 | 0.0 | 0.2 | 0.1 | 0.5 | 0.4 | 1.1 | 0.1 | 0.1 | 0.1 | 0.1 | 1.4 | |
| 26 | 0.0 | 0.0 | 0.1 | 0.1 | 0.4 | 0.3 | 4.0 | 0.1 | 0.0 | 0.1 | 0.1 | 0.7 | |
| 27 | 0.0 | 0.0 | 0.0 | 0.1 | 0.4 | 0.3 | 3.2 | 0.1 | 0.0 | 0.1 | 0.1 | 0.6 | |
| 28 | 0.0 | 0.0 | 0.0 | 0.1 | 0.3 | 0.3 | 1.3 | 0.1 | 0.0 | 0.5 | 0.1 | 0.6 | |
| 29 | 0.0 | 0.0 | 0.0 | 0.1 | 0.3 | 0.3 | 0.8 | 0.1 | 0.0 | 0.5 | 0.1 | 0.4 | |
| 30 | 0.0 | 0.0 | 0.0 | 0.1 | 0.3 | . . | 0.5 | 0.1 | 0.0 | 0.3 | 0.1 | 0.4 | |
| 31 | 0.0 | 0.0 | 0.0 | 0.1 | 0.3 | . . | 0.6 | . . | 0.0 | . . | 0.1 | 0.2 | |

NOTE — In the original monthly table, the stages given were the heights of the surface above the bottom of the gauge-rod, which point was 0.3 foot below the 0.0 of this table.

### FIG. 8

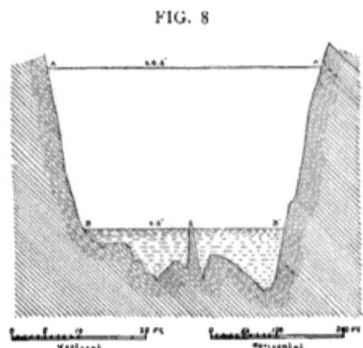

CROSS-SECTION OF THE OCMULGEE RIVER, AT MACON, GEORGIA

## FLOW OF THE OCMULGEE RIVER, AT MACON, GEORGIA

| No. | Date of Measurement | Stage | Area in Sq. Ft. of Cross-Section | Velocity in Feet per Second | Discharge in Cu. Ft. per Sec. | REMARKS |
|---|---|---|---|---|---|---|
| 1 | Aug. 18, 1891 | 3.2 | 2,444.7 | 1.47 | 3,611.7 | Measured by C. C. Anderson. |
| 2 | Nov. 28, 1892 | 20.6 | 5,800.0 | 4.35 | 25,269.6 | Measured by C. C. Anderson. |
| 3 | . . . . . . | 0.0 | . . . . . | . . . . | 2,157.6 | Stated in notes by C. C. Anderson. |

## TABLE VIII
### DAILY FLUCTUATIONS IN FEET AND TENTHS
*Lowest Observed Stage = 0.0*
### THE OCMULGEE RIVER, AT MACON, GEORGIA

| Date | 1891 | | | | | 1892 | | | | | | |
|---|---|---|---|---|---|---|---|---|---|---|---|---|
| | Aug. | Sept. | Oct. | Nov. | Dec. | Jan. | Feb. | Mar. | Apr. | May | June | July | Aug. |
| 1 | | 2.4 | 1.9 | 0.1 | | | | 5.0 | 3.8 | 3.7 | | | 1.8 |
| 2 | | 2.7 | 1.7 | 0.2 | | | | 4.9 | 3.9 | 3.7 | | | 2.4 |
| 3 | | 4.5 | 2.7 | 0.0 | | | | 4.6 | 3.8 | 3.6 | | | 2.4 |
| 4 | | 2.8 | 2.0 | 0.0 | | | 4.4 | 4.2 | 4.1 | 3.6 | | | 2.4 |
| 5 | | 4.8 | 1.6 | 0.6 | | | 4.2 | 4.1 | 4.1 | 3.7 | | No Record. | 8.2 |
| 6 | Begun August 18th, 1891. | 1.7 | 1.6 | 0.8 | . | | 4.1 | 3.8 | 4.2 | 3.6 | | | 7.0 |
| 7 | | 1.6 | 1.6 | 0.8 | | | 4.1 | 5.9 | 8.9 | 3.6 | | | 4.8 |
| 8 | | 2.5 | 1.7 | 0.9 | | | 6.8 | 9.0 | 15.0 | 3.9 | | | 4.6 |
| 9 | | 2.0 | 1.8 | 0.8 | | | 14.4 | 13.9 | 14.9 | 4.7 | | | 4.2 |
| 10 | | 1.8 | 1.8 | 1.6 | | | 13.0 | 9.8 | 14.6 | | | | 2.6 |
| 11 | | 1.7 | 1.9 | 1.9 | | | 8.8 | 6.9 | 12.8 | | | | 2.6 |
| 12 | | 1.8 | 1.8 | 2.2 | | | 7.0 | 4.8 | 9.1 | | | 11.6 | 6.6 |
| 13 | | 3.4 | 1.7 | 2.8 | | | 5.9 | 7.7 | 5.8 | | | 8.9 | 5.3 |
| 14 | | 3.3 | 1.7 | 2.0 | No Record. | No Record. | 5.2 | 6.8 | 4.7 | | No Record. | 6.0 | 4.2 |
| 15 | | 3.4 | 1.6 | 1.4 | | | 7.4 | 5.6 | 3.8 | | | 7.8 | 4.6 |
| 16 | | 2.4 | 1.6 | 1.7 | | | 7.6 | 5.6 | 4.6 | | | 5.0 | |
| 17 | | 2.7 | 0.6 | 1.8 | | | 7.2 | 5.0 | 4.2 | | | 4.6 | |
| 18 | 3.2 | 2.0 | 0.6 | 3.2 | | | 6.8 | 4.9 | 3.9 | | | 4.6 | |
| 19 | 2.8 | 1.8 | 0.6 | 3.3 | | | 5.0 | 4.9 | 3.9 | | | 4.4 | |
| 20 | 2.7 | 2.8 | 0.4 | 3.4 | | | 4.6 | 4.8 | 4.0 | | No Record. | 11.4 | Ended August 15th, 1892. |
| 21 | 3.2 | 1.7 | 0.4 | 3.8 | | | 11.2 | 4.6 | 4.0 | | | 8.0 | |
| 22 | 3.3 | 1.6 | 0.3 | 1.8 | | | 13.9 | 4.4 | 4.0 | | | 6.8 | |
| 23 | 6.3 | 2.6 | 0.2 | 1.8 | | | 11.4 | 4.7 | 3.7 | | | 6.2 | |
| 24 | 13.3 | 2.3 | 0.2 | 3.9 | | | 7.8 | 5.1 | 3.7 | | | 5.8 | |
| 25 | 11.6 | 2.8 | 0.2 | 5.2 | | | 7.2 | 14.3 | 3.6 | | | 5.4 | |
| 26 | 17.4 | 1.7 | 0.4 | 5.9 | | | 7.2 | 17.6 | 3.6 | | | 5.4 | |
| 27 | 15.3 | 1.7 | 0.5 | 5.9 | | | 5.2 | 24.6 | 3.6 | | | 3.6 | |
| 28 | 11.6 | 1.6 | 0.2 | 4.0 | | | 5.0 | 20.6 | 3.6 | | | 3.0 | |
| 29 | 6.8 | 1.6 | 0.1 | 2.0 | | | . . | 15.6 | 3.6 | | | 2.6 | |
| 30 | 5.3 | 1.6 | 0.0 | 2.0 | | | . . | 9.4 | 3.7 | | | 2.6 | |
| 31 | 3.4 | . . | 0.0 | . . | | | . . | 5.3 | . . | | | 2.4 | |

NOTE.— In the original monthly table, the stages given were the heights of the surface above the bottom of the gauge-rod, which point was 0.4 foot below the 0.0 of this table.

FIG. 9

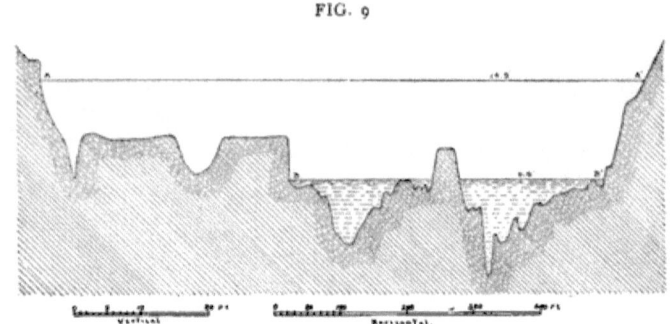

CROSS-SECTION OF THE OCMULGEE RIVER, AT JULIETTE, MONROE COUNTY, GEORGIA

## FLOW OF THE OCMULGEE RIVER, AT JULIETTE, MONROE COUNTY, GEORGIA

| No. | Date of Measurement | Stage | Area in Sq. Ft. of Cross-Section | Velocity in Feet per Second | Discharge in Cu. Ft. per Sec. | REMARKS |
|---|---|---|---|---|---|---|
| 1 | Sept. 4, 1891 | 0.5 | 2,300.5 | 1.57 | 3,615.6 | Measured by C. C. Anderson. |
| 2 | May 6, 1892 | 0.4 | 2,258.0 | 1.19 | 2,691.2 | Measured by C. C. Anderson. |

Data not sufficient for calculating minimum discharge.

## TABLE IX
### DAILY FLUCTUATIONS IN FEET AND TENTHS
*Lowest Observed Stage = 0.0*
THE OCMULGEE RIVER, AT JULIETTE, MONROE COUNTY, GEORGIA

| Date | 1891 | | | | 1892 | | | | | | | | |
|---|---|---|---|---|---|---|---|---|---|---|---|---|---|
| | Sept. | Oct. | Nov. | Dec. | Jan. | Feb. | Mar. | Apr. | May | June | July | Aug. | Sept. |
| 1 | Begun Sept. 4th, 1891. | 1.1 | 1.1 | 1.2 | 1.1 | 1.9 | 1.8 | 1.1 | 0.7 | 0.3 | 0.3 | 0.1 | 0.2 |
| 2 | | 1.1 | 1.1 | 1.2 | 1.3 | 1.9 | 1.8 | 1.0 | 0.6 | 0.3 | 0.3 | 0.3 | 0.2 |
| 3 | | 1.1 | 1.1 | 1.2 | 1.4 | 1.8 | 1.7 | 0.9 | 0.6 | 0.6 | 0.3 | 0.8 | 0.2 |
| 4 | 1.6 | 1.1 | 1.1 | 1.2 | 1.5 | 1.7 | 1.7 | 0.9 | 0.6 | 0.7 | 0.2 | 0.6 | 0.1 |
| 5 | 1.6 | 1.1 | 1.1 | 1.7 | 1.4 | 1.7 | 1.7 | 0.8 | 0.5 | 0.6 | 0.4 | 0.3 | 0.1 |
| 6 | 1.4 | 1.1 | 1.1 | 1.9 | 1.7 | 1.5 | 1.5 | 1.0 | 0.5 | 0.5 | 0.7 | 0.2 | 0.0 |
| 7 | 1.4 | 1.1 | 1.1 | 1.5 | 2.7 | 1.5 | 1.4 | 2.6 | 0.5 | 0.4 | 0.3 | 0.2 | 0.0 |
| 8 | 1.4 | 1.1 | 1.1 | 1.6 | 2.8 | . . . | 2.9 | 6.4 | 0.4 | 0.6 | 0.3 | 0.3 | 0.0 |
| 9 | 1.3 | 1.1 | 1.1 | 1.6 | 1.4 | 4.5 | 2.1 | 5.5 | 0.4 | 0.8 | 0.3 | 0.5 | |
| 10 | 1.2 | 1.1 | 1.1 | 1.6 | 1.4 | 2.9 | 1.5 | 1.9 | 0.9 | 0.9 | 2.1 | 0.4 | |
| 11 | 1.2 | 1.1 | 1.2 | 1.3 | 1.8 | 2.3 | 1.1 | 1.8 | 1.0 | 0.5 | 2.8 | 0.3 | |
| 12 | 1.4 | 1.1 | 1.2 | 1.3 | 3.1 | 2.1 | 0.6 | 1.6 | 0.9 | 0.4 | 2.3 | 0.3 | |
| 13 | 1.4 | 1.1 | 1.2 | 1.3 | 4.1 | 2.1 | 0.6 | 1.1 | 0.6 | 0.3 | 2.3 | 0.3 | |
| 14 | 1.4 | 1.1 | 1.1 | 1.3 | 8.6 | 2.1 | 0.5 | 1.0 | 0.6 | 0.3 | 1.6 | 0.0 | |
| 15 | 1.3 | 1.1 | 1.1 | 1.2 | 10.7 | 2.1 | 0.7 | 1.0 | 0.6 | 0.2 | 1.1 | 0.0 | |
| 16 | 1.2 | 1.1 | 1.1 | 1.3 | 6.1 | 2.5 | 0.7 | 0.9 | 0.5 | 0.2 | 0.9 | 0.1 | Ended September 8th, 1892. |
| 17 | 1.2 | 1.1 | 1.1 | 1.5 | 3.0 | 2.0 | 9.7 | 0.9 | 0.4 | 0.1 | 0.7 | 0.3 | |
| 18 | 1.2 | 1.1 | 1.1 | 1.5 | 5.0 | 2.1 | 0.9 | 0.8 | 0.5 | 0.2 | 1.1 | 2.5 | |
| 19 | 1.2 | 1.1 | 1.1 | 1.5 | 5.1 | 1.9 | 1.0 | 0.8 | 0.5 | 0.4 | 0.9 | 2.5 | |
| 20 | 1.2 | 1.1 | 1.1 | 1.4 | 14.9 | 1.7 | 0.8 | 0.8 | 0.5 | 0.8 | 2.2 | 1.8 | |
| 21 | 1.2 | 1.1 | 1.1 | 1.4 | 13.1 | 3.0 | 0.7 | 0.8 | 0.4 | 0.7 | 1.6 | 1.7 | |
| 22 | 1.2 | 1.1 | 1.1 | 1.5 | 5.1 | 3.5 | 0.7 | 0.8 | 0.4 | 1.1 | 0.9 | 1.6 | |
| 23 | 1.2 | 1.1 | 2.0 | 1.5 | 2.9 | 2.8 | 0.7 | 0.9 | 0.6 | 1.3 | 0.5 | 1.9 | |
| 24 | 1.2 | 1.1 | 2.2 | 1.5 | 2.6 | 2.7 | 1.5 | 0.9 | 0.4 | 0.8 | 0.3 | 2.1 | |
| 25 | 1.2 | 1.1 | 1.7 | 1.5 | 2.5 | 2.1 | 3.1 | 0.8 | 0.3 | 0.7 | 0.5 | 1.4 | |
| 26 | 1.2 | 1.1 | 1.3 | 1.5 | 2.4 | 2.0 | 11.1 | 0.8 | 0.3 | 1.5 | 0.5 | 1.3 | |
| 27 | 1.2 | 1.1 | 1.2 | 1.4 | 2.1 | 2.1 | 12.0 | 0.8 | 0.3 | 1.6 | 0.4 | 2.2 | |
| 28 | 1.2 | 1.1 | 1.2 | 1.4 | 1.9 | 2.0 | 10.9 | 0.8 | 0.3 | 2.6 | 0.3 | 0.9 | |
| 29 | 1.2 | 1.1 | 1.2 | 1.3 | 1.9 | 1.9 | 2.6 | 0.7 | 0.1 | 1.4 | 0.3 | 0.7 | |
| 30 | 1.2 | 1.1 | 1.2 | 1.2 | 1.9 | . . . | 1.9 | 0.7 | 0.1 | 0.5 | 0.2 | 0.5 | |
| 31 | . . . | 1.1 | . . . | 1.1 | 1.9 | . . . | 1.7 | . . . | 0.1 | . . . | 0.2 | 0.3 | |

NOTE — In the original monthly table, the stages given were the heights of the surface above the bottom of the gauge-rod, which point was 0.9 foot below the 0.0 of this table.

## PLANS AND PROFILE

In addition to the foregoing cross-sections and fluctuation tables, Mr. Anderson's notes contained thirty-two illustrations, showing plans and profiles of important water-powers, some of which are partially utilized. A description of each is given below, and such reference is made to his three books of notes, on file in the office of the State Geologist, as will enable those, particularly interested, to examine the plans and profiles.

*1st.* SOQUEE RIVER. *Profile of Porter Mills Shoals.* Book No. 2, page 39. These three shoals cover a fall of 90 feet, in a total length of 6,600 feet. Shoal No. 1 falls 14.4 feet in 800 feet. This is the upper Cotton Mill Shoal. From the foot of this shoal, the river is comparatively level, for 2,000 feet, to the head of Shoal No. 2, the Woolen Mill Shoal, which falls 45.2 feet in a distance of 1,150 feet, and has an additional fall, below the Woolen Mill wheel, of 14 feet, in a distance of 1,950 feet. Shoal No. 3 begins at this point, and falls 15 feet in 700 feet. Volume of stream at low water, 250 cubic-feet per second. Net horse-power utilized, 250.

*2nd.* CHATTAHOOCHEE RIVER. *Plan and profile* showing its junction with the Soquee and three miles below this point. Book No. 2, page 41. It includes Duncan, Carpenter's, Gearing, Fish-trap, and Bull shoals. Total fall, 38 feet in a distance of 13,200 feet.

*3rd.* CHATTAHOOCHEE RIVER. *Island Ford Shoal.* Book No. 2, page 64. *Plan, profile and section.* Fall of 5.4 feet in 1,100 feet, or a 10-foot fall in 4,500 feet, from the top of the shoal to Roswell bridge.

*4th.* VICKERY'S CREEK. *At Roswell, Ga.* Book No. 2, page 63. This shows the upper Cotton Mill, the lower Cotton Mill and the Laurel Mills. All the power is utilized.

*5th.* CHATTAHOOCHEE RIVER. *Bull Sluice Shoal,* in Fulton and

WATER POWERS OF GEORGIA.  PLATE VIII

CANE CREEK FALLS, NEAR DAHLONEGA, GEORGIA.

Cobb counties. Book No. 2, page 65. A fall of 44 feet from Roswell bridge to the foot of Bull sluice. Distance 18,000 feet.

*6th.* CHATTAHOOCHEE RIVER. *Cochran Shoal* and *Devil's Racecourse*, Fulton and Cobb counties. Book No. 2, page 66. Fall, 17 feet in 8,000 feet.

*7th.* FLINT RIVER. *Flat Shoals*, Pike and Meriwether counties. Book No. 2, page 43. Fall, 32 feet in 3,000 feet.

*8th.* FLINT RIVER. *Dripping Rock Shoal*, Upson county. Book No. 2, page 58. A fall of 14 feet in 3,900 feet.

*9th.* FLINT RIVER. *Yellow Jacket Shoals*, Upson county. Book No. 2, page 53. A fall of 36.6 feet in 3,400 feet.

*10th.* FLINT RIVER. *Snipes Shoals*, Upson county. Book No. 2, page 60. A fall of 12 feet in 2,350 feet.

*11th.* BIG POTATO CREEK. *Rogers Shoal*, Upson county. Book No. 2, page 44. A fall of 80 feet in 3,600 feet.

*12th.* BIG POTATO CREEK. *Daniels Mill*, Upson county. Book No. 2, page 59. A fall of 13 feet in 150 feet.

*13th.* OCMULGEE RIVER. *Barnes Shoals*, at the junction of Yellow river and South river, Newton county. Book No. 2, page 52. A fall of 14 feet in 1,200 feet.

*14th.* OCMULGEE RIVER. *Key's Ferry*, Butts county. Book No. 2, page 75. A fall of 7.5 feet in 1,900 feet.

*15th.* OCMULGEE RIVER. *Pittman Ferry and Harper Shoals*, Butts county. Book No. 2, page 74. Falls 28 feet in 5,500 feet, at Harper Shoal, and 6 feet in 1,600 feet, below ferry.

*16th.* OCMULGEE RIVER. *Smith's Ferry and Lamar's Mill*, Butts county. Book No. 2, pages 45 and 67. A fall of 28 feet in 4,700 feet; at Lamar's mill, the fall is 18 feet in 1,000 feet.

*17th.* OCMULGEE RIVER. *Carden Shoal*, Monroe county. Book No. 2, page 62. A fall of 9 feet in 4,500 feet.

*18th*. OCMULGEE RIVER. *Holton*, Bibb county. Book No. 2, page *78*. A fall of 7 feet in 2,000 feet.

*19th*. YELLOW RIVER. *Porter Dale Mills, at Cedar Shoals*, Newton county. Book No. 2, page 49. Falls 54.7 feet in 2,200 feet.

*20th*. YELLOW RIVER. *Indian Fishery*, Newton county. Book No. 2, page 51. Falls 12 feet in 550 feet.

*21st*. SOUTH RIVER. *Snapping Shoals*, Newton county. Book No. 2, page 50. A fall of 28 feet in 1,500 feet.

*22nd*. ALCOVY RIVER. *Newton Factory on White and Garner Shoals*, Newton county. Book No. 2, page 48. A fall of 85 feet in 3,800 feet.

*23rd*. TOWALIGA RIVER. *High Falls*, Monroe county. Book No. 2, page 47. A fall of 95 feet in 600 feet.

*24th*. NORTH OCONEE RIVER. *Hurricane Shoals*, Jackson county. Book No. 2, page 68. Falls 30 feet in 600 feet.

*25th*. NORTH OCONEE RIVER. *Tumbling Shoals*, Jackson county. Book No. 2, page 72. Falls 8 feet in 600 feet.

*26th*. MIDDLE OCONEE RIVER. *Tallassee Bridge Shoal*, Jackson county. Book No. 2, page 77. Falls 31 feet in 3,600 feet.

*27th*. NORTH OCONEE RIVER. *Georgia Factory Shoal*, Clarke county. Book No. 2, page 97. A fall of 21 feet in 2,100 feet.

*28th*. MIDDLE OCONEE RIVER. *McElroy's Mill*, Clarke county. Book No. 2, page 81. A fall of 23 feet in 2,600 feet.

*29th*. MIDDLE OCONEE RIVER. *Princeton Factory*, Clarke county. Book No. 2, page 86. Falls 15 feet.

*30th*. OCONEE RIVER. *Barnett's Shoal*, Oconee county. Book No. 2, page 99. Falls 54 feet in 3,950 feet.

*31st*. APALACHEE RIVER. *High Shoals*, Oconee county. Book No. 2, page 98. Falls 50 feet in 600 feet.

*32nd*. APALACHEE RIVER. *Price's Shoal*, Oconee county. Book No. 2, page 100, and Book No. 3, page 32. A fall of 19 feet in 900 feet.

# CHAPTER V

## ELEVATIONS ON RAILROAD LINES

These tables were compiled by Mr. C. C. Anderson, C.E., late Assistant Geologist of this Survey. The following is an extract from his report concerning them:—

"These elevations for Topography were obtained from various railroads; but the list is by no means complete. Through the courtesy of the Chief Engineers of the Georgia Pacific, East Tennessee, Virginia and Georgia, Georgia Midland and Gulf, the Atlanta and Florida, and of the Assistant Engineers of the Central of Georgia System, Georgia, Southern and Florida, and the Savannah, Florida and Western, a list of the elevations of the various mile-posts and railroad stations has been obtained and reported. Some of these elevations refer to cross-ties or grade, while others refer to ground surface. At the tie-points, where the roads meet, or cross each other, it has been found impossible to harmonize the datum lines of the respective roads, for the reason, that no fixed points have been determined, from which to make the ties. This has been especially difficult at Macon, where some level-notes refer to surface, and others, to grade. The notes of the S., F. and W. were complete and accurate from Savannah to Bainbridge and from Waycross to Albany So are those of the Central, when the datum for such was taken from mean low tide at Savannah. The U. S. Coast Survey has made a change of this datum, from mean low tide at Savannah to mean low tide at Fort Pulaski, where daily readings have been kept up, for a number of years. To this datum have all elevations been reduced, where possible.

It is necessary to mention the grave discrepancy in the elevation at the car-shed in Atlanta, as given by the level-notes of the Central R. R., and those of the U. S. Coast and Geodetic Survey. The Central R. R. notes show Atlanta to be 1,085 feet above mean low tide at Fort Pulaski, while the Geodetic Survey shows the elevation to be 1,050 feet above the same datum.

Mr. Schwab, Assistant Engineer and Draughtsman to the Central R. R., at Savannah, through whose courtesy these notes were obtained, has reduced all the lines of the Central System to one common datum of the main line at Savannah, which is given as zero. This zero-point is forty-six feet above mean low tide at Fort Pulaski, as found by Mr. Geisler, Assistant Engineer of the Coast Survey, who established permanent benches, or "B. M.," at various points in Savannah, as points of reference. From one of these bench-marks, levels were run to the head of the track in the Central passenger-depot, in that city, with the above result; that is, the Central datum to be forty-six feet above mean low tide at Fort Pulaski.

Mr. Schwab has carefully corrected, compiled and reduced all the levels of the Central System to this zero datum, with the result of making Atlanta 1,085 feet, instead of 1,050 feet. How this elevation of 1,050 feet was ever determined is not known. Mr. Schwab's figures are relied on for accuracy. His long years of experience; his familiarity with the Central R. R. notes, field and office; his known exact methods of work, give credit to the assumption, that 1,085 feet is the correct elevation for Atlanta.

The Southwestern R. R. and the B. & W. R. R., the former from Macon and the latter from Brunswick, meet at Albany, where the two different routes from Savannah harmonize very closely. This is close enough to give confidence to the levels, as run and worked out by the two routes.

The datum of the Georgia, Southern & Florida, which starts at Macon, was assumed at 200 feet, when the preliminary survey was made. This datum was retained, during location and construction. It crosses the B. & W. at Tifton. An attempt has been made to harmonize the levels at this point; but not very successfully, on account of the notes of the B. & W. referring to ground surface. When the elevations of the B. & W. station at Tifton are applied to the Ga., Sou. & Fla. at the same point, the Ga., Sou. & Fla. elevations, at the Union passenger-depot in Macon, do not correspond with the Central R. R. elevations, at the same point. This discrepancy can be reconciled, if the points, to which the elevations of either road refer, can be located and fixed with exactness. The importance of these railroad elevations cannot be overestimated; as so many topographical and geological questions depend upon them."

# TIDE-WATER ELEVATIONS ON RAILROAD LINES

### COMPILED BY C. C. ANDERSON
Ex-Assistant State Geologist

| GEORGIA MIDLAND & GULF R. R.[1] | | COLUMBUS SOUTHERN R. R. | |
|---|---|---|---|
| Station | Elevation[2] | Station | Elevation[2] |
| Columbus | 260.0 | Columbus | 260.0 |
| Flat Rock | 474.0 | Bull Creek | 240.0 |
| Bull Creek | 408.0 | Upatoie | 225.0 |
| Midland | 565.0 | Ochillee | 289.0 |
| Ellerslie | 726.0 | Cusseta | 532.0 |
| Waverly Hall | 746.0 | Manta | 515.0 |
| Mulberry Creek | 632.0 | Top of Cut, Manta | 565.0 |
| Mulberry Oak Mountain | 716.0 | Green Hill | 601.0 |
| Shiloh | 919.0 | Brooklyn | 691.0 |
| Tennille | 1,060.0 | Richland | 600.0 |
| Topover Mountain, over Tunnel | 1,148.0 | Westerio | 528.0 |
| Nebula | 1,039.0 | Parrott's | 482.0 |
| Warm Springs | 929.0 | Dawson | 376.0 |
| Cold Creek | 753.0 | Sasser | 336.0 |
| Raleigh | 765.0 | Oakland | 275.0 |
| Cane Creek | 705.0 | Palmyra | 260.0 |
| Woodbury | 781.0 | Albany | 208.0 |
| Flint River | 658.0 | | |
| Molena | 780.0 | EAST TENNESSEE, VIRGINIA & GEORGIA R. R.[1] | |
| Neal | 824.0 | Station | Elevation |
| Concord | 820.0 | | |
| Williamson's | 931.0 | Red Clay | 841.0 |
| Griffin | 967.0 | Cohutta | 885.0 |
| Towaliga River | 682.0 | Varnell's | 824.0 |
| Lowella | 871.0 | Waring's | 813.0 |
| Greenwood | 873.0 | Dalton | 775.0 |
| McDonough | 870.0 | Immiline Creek | 708.0 |

[1] Now a part of the Southern Railway.  [2] In feet.

## RAILROAD ELEVATIONS

### EAST TENNESSEE, VIRGINIA & GEORGIA R. R.[1]
(*Continued*)

| Station | Elevation[2] |
|---|---|
| Phelps | 724.0 |
| Carbondale | 776.0 |
| Miller's | 731.0 |
| Valley | 661.0 |
| Snake Creek | 630.0 |
| Bruse Creek | 624.0 |
| Oostanaula | 646.0 |
| Bottom | 597.0 |
| Oostanaula River | 620.0 |
| Creek, 90 Mile-post | 658.0 |
| Plainville | 690.0 |
| Shannon | 698.0 |
| Harper | 691.0 |
| Stream | 659.0 |
| Creek, 79 Mile-post | 679.0 |
| North Rome | 643.0 |
| Etowah River | 635.0 |
| East Rome | 624.0 |
| Atlanta Junction | 619.0 |
| Vance Creek | 614.0 |
| Silver Creek, 73 Mile-post | 612.0 |
| Silver Creek, 70 Mile-post | 688.0 |
| Dry Creek | 793.0 |
| Seney | 842.0 |
| Fish Creek | 755.0 |
| Euharlee | 776.0 |
| Rockmart | 775.0 |
| Braswell | 1,066.0 |
| Summit | 1,200.0 |
| Big Tunnel | 1,095.0 |
| Stream | 937.0 |
| Cochrane Creek | 1,012.0 |
| Stream | 908.0 |
| Big Raccoon Creek | 988.0 |
| Top of Summit | 1,073.0 |

### EAST TENNESSEE, VIRGINIA & GEORGIA R. R.[1]
(*Continued*)

| Station | Elevation[2] |
|---|---|
| Little Tunnel | 1,002.0 |
| Little Raccoon Creek | 981.0 |
| McPherson | 1,011.0 |
| Stream | 856.0 |
| Pumpkinvine Creek | 920.0 |
| Dallas | 1,012.0 |
| Big Powder Springs Creek | 927.0 |
| Powder Springs | 921.0 |
| Stream | 889.0 |
| Sweet Water Creek | 904.0 |
| Austell | 935.0 |
| Peters Street, Atlanta | 1,054.0 |
| Railroad Shops, Atlanta | 1,028.0 |
| Atlanta | 1,038.0 |
| Stream | 784.0 |
| South River | 826.0 |
| Summit | 932.0 |
| Soapstone Cut | 905.0 |
| Stream | 816.0 |
| Creek | 861.0 |
| Ellenwood | 853.0 |
| Estes | 768.0 |
| Stream | 736.0 |
| Indian Creek | 751.0 |
| Stockbridge | 803.0 |
| Indian River | 714.0 |
| Pates' Creek | 714.0 |
| Stream | 741.0 |
| Walnut Creek | 791.0 |
| Stream | 748.0 |
| Camp Creek | 778.0 |
| Long Branch | 778.0 |
| McDonough | 852.0 |
| Near McDonough | 890.0 |
| Cloud's Branch | 873.9 |

[1] Now a part of the Southern Railway.  [2] In feet.

## RAILROAD ELEVATIONS

### EAST TENNESSEE, VIRGINIA & GEORGIA R. R.[1]
*(Continued)*

| Station | Elevation[2] |
|---|---|
| Locust Grove | 825.0 |
| Yellow Water Creek | 662.0 |
| Jackson | 705.0 |
| Flovilla | 655.0 |
| Williams | 632.0 |
| Stream | 385.0 |
| Big Sandy | 410.0 |
| Stream | 385.0 |
| Rattlesnake Creek | 408.0 |
| Stream | 380.0 |
| Towaliga River | 415.0 |
| Juliette | 395.0 |
| Stream | 354.0 |
| Powder Creek | 384.0 |
| Dames Ferry | 364.0 |
| Stream | 327.0 |
| Rum Creek | 353.0 |
| Holton | 350.0 |
| Stream | 300.0 |
| Beaver Creek | 321.0 |
| Stream | 283.0 |
| Vineville Branch | 303.0 |
| Macon | 311.0 |
| Cotton Yard | 311.0 |
| Stratton's Branch | 285.0 |
| Banks of Stream | 272.0 |
| Ocmulgee River | 285.0 |
| Reid's | 280.0 |
| Bullard's | 265.0 |
| Belchers Branch | 258.0 |
| Adams' Park | 265.0 |
| Savage Creek | 251.0 |
| West Lake | 240.0 |
| Coley's | 306.0 |
| Cochrane | 341.0 |

### EAST TENNESSEE, VIRGINIA & GEORGIA R. R.[1]
*(Continued)*

| Station | Elevation[2] |
|---|---|
| Empire | 380.0 |
| Dubois | 394.0 |
| Dempsey | 363.0 |
| Eastman | 361.0 |
| Godwinsville | 316.0 |
| Chancey | 303.0 |
| Cunningham | 707.0 |
| Cave Spring | 697.0 |
| State Line | 900.0 |

### GEORGIA PACIFIC R. R.[1]

| Station | Elevation[2] |
|---|---|
| Peyton | 870.0 |
| Chattahoochee River | 808.0 |
| Nickajack Creek | 808.0 |
| Stanback's Creek | 820.0 |
| Nickajack No. 2 | 839.0 |
| " No. 3 | 850.0 |
| Mable's Trestle | 922.0 |
| Near Mableton | 1,006.0 |
| Mableton | 986.0 |
| Water Tank | 936.0 |
| Sweetwater | 913.0 |
| Austell | 937.0 |
| County Line | 1,010.0 |
| Salt Springs | 1,034.0 |
| Douglasville | 1,216.0 |
| Winston | 1,130.0 |
| County Line | 1,146.0 |
| Villa Rica | 1,157.0 |
| Water Tank, 40 Mile-post | 1,054.0 |
| Tallapoosa River, Little | 1,057.0 |
| Temple | 1,178.0 |
| County Line | 1,221.0 |

[1] Now a part of the Southern Railway.  [2] In feet.

TOCCOA FALLS, HABERSHAM COUNTY, GEORGIA.

## GEORGIA PACIFIC R. R.[1]
### (Continued)

| Station | Elevation [2] |
|---|---|
| Bremen | 1,413.0 |
| Waco | 1,343.0 |
| Tallapoosa | 1,159.0 |
| Tallapoosa River | 963.0 |
| Dempsey Creek | 943.0 |
| State Line | 945.0 |

## SAVANNAH, FLORIDA & WESTERN R. R.

| Station | Elevation [2] |
|---|---|
| Savannah | 25.8 |
| Little Ogeechee River | 19.0 |
| Crosstie, East End of Bridge | 18.4 |
| "    West  "   " | 17.7 |
| Bottom of River | 9.8 |
| Station, No. 10 | 25.8 |
| Burroughs, No. 12 | 17.8 |
| Great Ogeechee Bridge | 20.6 |
| Bottom of River | 9.8 |
| Station, No. 16, or Way's | 21.1 |
| Branch, 18½ Miles | 22.2 |
| Branch, 20½ Miles | 22.6 |
| Branch, 21¼ Miles | 22.5 |
| Mt. Hope Creek | 23.3 |
| Branch, 22½ Miles | 24.2 |
| Flemming, No. 24 | 25.4 |
| Branch, 25 Mile-post | 23.0 |
| "    26½ Miles | 22.5 |
| "    27½  " | 22.3 |
| Branch, 29 Mile-post | 22.1 |
| "    29¼ Miles | 22.3 |
| McIntosh, No. 31 | 26.4 |
| McIntosh Creek | 22.8 |
| Gauldin's Creek | 29.6 |
| Branch | 31.9 |

## SAVANNAH, FLORIDA & WESTERN R. R.
### (Continued)

| Station | Elevation [2] |
|---|---|
| Branch, 38¼ Miles | 102.6 |
| Walthourville, No. 39 | 102.5 |
| Branch | 102.3 |
| " | 102.2 |
| " | 91.8 |
| "    40¼ Miles | 89.7 |
| "    41¾  " | 74.3 |
| Durham Creek | 66.3 |
| Johnston, No. 46 | 75.8 |
| Jones Creek | 52.6 |
| Fountain Branch | 50.8 |
| Forest Pond | 51.2 |
| Morgan Lake | 51.7 |
| Bottom of Lake | 12.8 |
| Water-surface of Lake | 38.3 |
| Altamaha River | 75.9 |
| High-water Mark | 42.5 |
| Mean-water Surface | 33.3 |
| Bottom of River | 21.3 |
| Doctortown, No. 53 | 77.3 |
| End of Cut | 92.1 |
| Jesup | 102.9 |
| "   Warehouse | 102.8 |
| Turnout, No. 62 | 101.7 |
| Dale's Mill, No. 67 | 140.0 |
| Branch, 67½ Miles | 126.8 |
| "    68¼  " | 121.8 |
| Screven, No. 69 | 127.3 |
| Turnout, No. 74 | 76.5 |
| Offerman, No. 76 | 110.4 |
| Patterson, No. 79 | 108.0 |
| Turnout, No. 83 | 127.8 |
| Blackshear, No. 87 | 125.8 |
| Turnout, No. 89 | 141.0 |
| Exeter, No. 93 | 96.8 |

[1] Now a part of the Southern Railway.    [2] In feet.

## SAVANNAH, FLORIDA & WESTERN R. R.
### (Continued)

| Station | Elevation[1] |
|---|---|
| Big Satilla River | 96.4 |
| Bottom of River | 68.8 |
| Water-surface of River | 71.8 |
| Waycross, No. 97 | 140.8 |
| Turnout, No. 99 | 147.1 |
| Glenmore's, No. 103 | 112.1 |
| Argyle, No. 116 | 164.2 |
| Homerville, No. 123 | 179.8 |
| Dupont, No. 131 | 184.1 |
| Junction, No. 131 | 184.1 |
| Stockton, No. 139 | 192.6 |
| Naylor | 195.6 |
| Valdosta | 218.8 |
| Ousley | 151.8 |
| Quitman | 176.7 |
| Dixie | 134.4 |
| Boston | 197.9 |
| Thomasville | 253.6 |
| Cairo | 242.4 |
| Whigham | 268.9 |
| Climax | 280.8 |
| Bainbridge | 113.6 |
| Fowltown | 292.8 |
| Franceville | 299.8 |
| Recovery | 192.8 |
| Florida Railway & Navigation Co. | 75.8 |
| Chattahoochee | 73.8 |
| Pensacola Junction | 74.8 |

## WAYCROSS & JACKSONVILLE BRANCH, SAVANNAH, FLORIDA & WESTERN R. R.

| Station | Elevation[1] |
|---|---|
| Waycross | 140.8 |
| Braganza | 147.8 |
| Fort Mudge | 137.8 |

## WAYCROSS & JACKSONVILLE BRANCH, SAVANNAH, FLORIDA & WESTERN R. R.
### (Continued)

| Station | Elevation[1] |
|---|---|
| Race Pond | 151.8 |
| Uptonsville | 87.3 |
| Folkstone | 83.8 |
| Boulogne | 73.8 |

## BRUNSWICK & WESTERN R. R.

| Station | Elevation[1] |
|---|---|
| Brunswick | 17.8 |
| Buffalo Swamp | 7.8 |
| Water Surface, Big Buffalo | 3.8 |
| Water Surface, Little Buffalo | 3.8 |
| Near Waynesville | 53.8 |
| Satilla River | 18.8 |
| Caney Bay | 103.8 |
| Big Creek, Water Surface | 80.8 |
| Waycross | 140.8 |
| Cox Creek | 104.8 |
| Waresboro | 120.8 |
| Dixonia Station | 126.8 |
| Poley Branch, Water Surface | 123.8 |
| Peach Creek, Water Surface | 94.8 |
| Gordonia | 131.8 |
| Duncan Branch, Water Surface | 117.8 |
| Red Bluff Creek | 108.3 |
| Branch, Red Bluff Station | 147.3 |
| Pearson Station | 172.8 |
| Kirkland | 200.8 |
| Westonia | 196.8 |
| Leliaton | 203.8 |
| Branch at 99 Mile-post | 196.3 |
| Pine Bloom | 206.8 |
| Willacoochee | 202.8 |
| Branch, 103 Mile-post | 176.8 |

[1] In feet.

## BRUNSWICK & WESTERN R. R.
*(Continued)*

| Station | Elevation[1] |
|---|---|
| Willacoochee River | 184.8 |
| " " | 179.3 |
| Sniff Station | 223.8 |
| Allapaha River | 121.8 |
| Branch of the Allapaha River | 241.8 |
| Allapaha Station | 268.8 |
| Branch of the Willacoochee River | 263.8 |
| Branch of the Willacoochee River | 259.8 |
| Ridge, 116 Mile-post | 388.8 |
| Enigma Station | 265.8 |
| Henry's Branch, 119¼ Miles | 248.8 |
| Brookfield | 306.8 |
| Middle Creek | 278.8 |
| New River | 282.8 |
| Vanceville | 290.8 |
| Little River | 303.8 |
| Tifton | 343.8 |
| Branch, 130 Mile-post | 304.8 |
| Tucker Creek | 255.8 |
| Riverside Station | 264.8 |
| Little River | 239.8 |
| Hillsdale Station | 303.8 |
| Ty Ty Creek | 275.8 |
| " " " and Station | 269.8 |
| Sumner Station | 350.8 |
| Wiston Mill | 351.8 |
| Poulan Station | 312.8 |
| Warrior Creek | 302.8 |
| Hog-heaven | 331.8 |
| Isabella | 341.8 |
| Coleman's Station | 354.8 |
| Willingham Station | 299.8 |
| Acrosta Station | 205.0 |

## BRUNSWICK & WESTERN R. R.
*(Continued)*

| Station | Elevation[1] |
|---|---|
| East Albany | 186.0 |
| Flint River Valley | 154.0 |
| Water Surface, Flint River | 127.0 |
| Albany | 172.0 |

## EAST GEORGIA & FLORIDA R. R.[2]

| Station | Elevation[1] |
|---|---|
| Jesup | 103.0 |
| Cypress Flat | 104.0 |
| Pigeon Roost Swamp | 95.0 |
| Branch, 63 Mile-post | 85.0 |
| Buffalo Creek | 66.0 |
| Crossing East Tenn., Va., & Ga. R. R.[3] | 68.0 |
| Turkey Swamp, 72 Mile-post | 75.0 |
| " " 72½ Miles | 67.0 |
| B. & W. R. R. Grade | 73.0 |
| College Creek | 63.0 |
| Little Satilla Swamp | 61.0 |
| Waverly Swamp | 60.0 |
| White Oak Swamp | 60.0 |
| Flowers' Swamp | 56.0 |
| Big Walker Swamp | 61.0 |
| Little Walker Swamp | 62.0 |
| Rose Creek Swamp | 71.0 |
| Seal Swamp | 61.0 |
| North Fork of Crooked River Swamp | 58.0 |
| Crooked River Bottom | 56.0 |
| South Fork, Crooked River Bottom | 55.0 |
| Little Catfish Creek | 56.0 |
| Big Catfish Creek | 44.0 |
| St. Mary's Swamp | 47.0 |
| St. Mary's River, Low Tide | 52.0 |

[1] In feet.
[2] Datum :— Reduced to Fort Pulaski by adding.
[3] Now a part of the Southern Railway.

## WESTERN & ATLANTIC R. R.[1]

| Station | Elevation[2] |
|---|---|
| Atlanta | 1,050.0 |
| Simpson Street Crossing | 1,025.6 |
| Belt Crossing | 969.7 |
| Guano Works | 937.8 |
| Chemical Works | 925.7 |
| Bolton | 848.3 |
| Iceville | 843.3 |
| Joplin | 837.6 |
| Collins Brick-yard | 851.6 |
| Chattahoochee River, crosstie | 833.0 |
| Gilmore | 900.2 |
| Vining's Station | 945.7 |
| McIver's | 967.0 |
| Smyrna | 1,068.4 |
| Ruff's | 1,065.6 |
| Marietta | 1,133.4 |
| Elizabeth | 1,164.4 |
| Big Shanty | 1,107.8 |
| Acworth | 929.0 |
| County Line | 910.1 |
| Allatoona Creek | 877.7 |
| Allatoona Station | 879.6 |
| Forty-one Junction | 871.2 |
| Bartow | 847.8 |
| Emerson | 843.7 |
| Etowah Junction | 755.8 |
| Etowah River | 746.0 |
| Cartersville | 762.2 |
| East & West Railroad Junction | 748.0 |
| Stream, 493 Mile-post | 731.9 |
| Rogers' R. R. Junction | 740.0 |
| Stream No. 40 | 744.0 |
| " " 39 | 754.0 |
| " " 38 | 758.9 |
| Cassville | 767.6 |

## WESTERN & ATLANTIC R. R.[1]
### (Continued)

| Station | Elevation[2] |
|---|---|
| Stream No. 37 | 782.3 |
| " " 36 | 759.9 |
| Best's | 750.0 |
| Gaines' Mill | 730.8 |
| Two Run Creek, No. 35 | 729.6 |
| Kingston | 712.7 |
| Cement | 687.3 |
| Hall's | 787.8 |
| Summit | 800.1 |
| Top of Grade | 808.2 |
| Oothcaloga River | 734.3 |
| Stream, 68 Mile-post | 708.3 |
| Adairsville | 722.1 |
| Oothcaloga River | 682.0 |
| County Line | 679.0 |
| McDaniel's | 669.9 |
| Oothcaloga River | 645.2 |
| Calhoun | 660.6 |
| Resaca | 657.5 |
| Oostanaula River | 657.4 |
| County Line | 659.3 |
| Tilton | 668.2 |
| Beardsley | 668.3 |
| Stream No. 24 | 727.1 |
| Dalton | 773.2 |
| Rock Face | 782.2 |
| 104 Mile-post | 783.3 |
| Tunnel Hill | 850.8 |
| County Line | 823.7 |
| Greenwood | 794.0 |
| Catoosa | 789.2 |
| Ringgold | 794.5 |
| Graysville | 711.0 |
| State Line | 715.0 |

[1] Datum:— Atlanta elevation, Union Depot, 1,050 feet above sea-level.   [2] In feet.

## GEORGIA, SOUTHERN & FLORIDA R. R.

| Station | Elevation [1] |
|---|---|
| Station O | 304.0 |
| Switch | 344.0 |
| Southwestern Railroad | 337.0 |
| Macon & Birmingham Railroad | 321.0 |
| River Swamp, North Edge | 287.0 |
| " " proper | 283.0 |
| " " proper | 278.0 |
| Macon & Birmingham Railroad | 279.0 |
| Last Lake | 278.0 |
| Ridge between River and Tobesofkee Creek | 309.0 |
| Creek Swamp | 277.0 |
| Ridge between Echeconnee Creek and Tobesofkee Creek | 332.0 |
| Ridge, Section-house, No. 7 | 363.0 |
| Ridge, Section-house, No. 8 | 289.0 |
| Avondale | 339.0 |
| Echeconnee Creek | 253.0 |
| Section-house, No. 14 | 298.0 |
| Joe Frederick | 286.0 |
| Willston, No. 16 | 295.0 |
| Sandy Reed Creek | 280.0 |
| Mrs. McBride's, No 10 | 331.0 |
| Section-house, No. 20 | 317.0 |
| Ridge, 20½ Miles | 344.0 |
| Beaver Creek | 292.0 |
| Ridge, 23½ Miles | 319.0 |
| Sofkee Junction | 335.0 |
| Kathleen | 318.0 |
| Section-house, No. 26 | 290.0 |
| Mossy Creek | 258.0 |
| Ridge between Big Indian and Mossy Creeks | 288.0 |
| Big Indian Creek | 294.0 |
| Limestone Creek | 294.0 |
| Hayneville Road | 311.0 |

## GEORGIA, SOUTHERN & FLORIDA R. R. (*Continued*)

| Station | Elevation [1] |
|---|---|
| Section-house, No. 35 | 421.0 |
| Top of Ridge, 35½ Miles | 451.0 |
| Holton Creek | 400.0 |
| Ridge, 38 Mile-post | 426.0 |
| Hawkinsville & Henderson Road | 413.0 |
| Big Creek | 311.0 |
| Ridge, 42½ Miles | 410.0 |
| John Croupler | 400.0 |
| Sub-grade, Macon & B. R'w'y | 321.0 |
| Section-house, No. 47 | 365.0 |
| Fullington Mill | 365.0 |
| Vienna | 319.0 |
| Section-house, No. 58 | 336.0 |
| Carnes Mill, 59½ Miles | 342.0 |
| Carnes Mill, 61¼ Miles | 359.0 |
| Savannah, Americus & Montgomery R. R. Crossing | 361.0 |
| Cordele | 388.0 |
| Section-house, No. 67 | 375.0 |
| Wenona, No. 69 | 394.0 |
| Vinton, No. 70 | 400.0 |
| Grady (?) Brown Place | 443.0 |
| Arabi Station | 399.0 |
| James's Saw-mill | 398.0 |
| Bedgood & Ryan | 404.0 |
| Pate's House | 396.0 |
| Section-house, No. 80 | 408.0 |
| Deep Creek | 350.0 |
| Section-house, No. 81 | 384.0 |
| Peckville | 446.0 |
| Marion, No. 85 | 451.0 |
| Branch, 86½ Miles | 409.0 |
| Sycamore | 397.0 |
| Inaha Station | 417.0 |
| Bottom, 92 Mile-post | 396.0 |

[1] In feet.

## GEORGIA, SOUTHERN & FLORIDA R. R.
### (Continued)

| Station | Elevation [1] |
|---|---|
| Brisham Road-grade | 405.0 |
| Cyclonetta | 413.0 |
| Wolf Pit | 394.0 |
| Section-house, No. 101 | 410.0 |
| "      "      " 102 | 415.0 |
| Brunswick & Western R. R. Crossing | 373.0 |
| Tifton Depot | 379.0 |
| Branch, 109½ Miles | 361.0 |
| Branch, 112¼ Miles | 336.0 |
| Hawell Mill | 301.0 |
| Laconte Station | 307.0 |
| 120 Mile-post | 272.0 |
| 121 Mile-post | 276.0 |
| Saw-mill and Still | 275.0 |
| 122 Mile-post | 273.0 |
| 123 Mile-post | 276.0 |
| Cypress Pond | 261.0 |
| Mill, 124½ Miles | 247.0 |
| Section-house, No 125 | 253.0 |
| Sparks Station | 244.0 |
| Troupville Road | 246.0 |
| Turkey Creek | 241.0 |
| 127 Mile-post | 249.0 |
| Adel Station | 252.0 |
| 129 Mile-post | 248.0 |
| 130 Mile-post | 240.0 |
| 131 Mile-post | 246.0 |
| Oxmoor Station | 252.0 |
| 135 Mile-post | 232.0 |
| 136 Mile-post | 229.0 |
| 137 Mile-post | 221.0 |
| 138 Mile-post | 236.0 |
| Vicker's Creek | 211.0 |
| Withlacoochee River | 140.0 |

## GEORGIA, SOUTHERN & FLORIDA R. R.
### (Continued)

| Station | Elevation [1] |
|---|---|
| Water-surface | 124.0 |
| Savannah, Florida & Western R. R. Crossing at Valdosta | 219.0 |
| Florida Midland R. R. | 209.0 |
| Center of Road-bed | 205.0 |
| Mike Bay | 204.0 |
| Mud Creek | 176.0 |
| 154 Mile-post | 203.0 |
| 155 Mile-post | 204.0 |
| 156 Mile-post | 190.0 |
| 157 Mile-post | 182.0 |
| Ulner's Mill | 200.0 |
| Long Pond | 180.0 |
| Lake Park | 167.0 |
| 164 Mile-post | 157.0 |
| Wessenboke House | 156.0 |
| State Line | 161.0 |
| Tank, 171 Mile-post | 151.0 |
| Allapaha River | 101.0 |
| 172 Mile-post | 105.0 |

## CENTRAL OF GEORGIA R. R.

| Station | Elevation [1] |
|---|---|
| Savannah | 46.0 |
| Junction, Meldrim | 39.3 |
| Egypt | 143.0 |
| Oliver | 140.0 |
| Little Ogeechee | 107.0 |
| Halcyondale | 112.0 |
| Outland | 110.0 |
| Ogeechee Station | 117.0 |
| Horse Creek | 136.0 |
| Scarboro Station | 157.0 |
| Paramonis Hill | 244.0 |

[1] In feet.

## RAILROAD ELEVATIONS

### CENTRAL OF GEORGIA R. R.
*(Continued)*

| Station | Elevation[1] |
|---|---|
| Ocains Branch | 199.0 |
| Ridge, 77 Mile-post | 210.0 |
| Millen Junction | 156.0 |
| Buckhead Creek | 156.0 |
| Rogers | 162.0 |
| Herndon | 189.0 |
| Sebastopol | 201.0 |
| Point, 98 Mile-post | 207.0 |
| Ogeechee River | 205.0 |
| Wadley Station | 243.0 |
| Bartow Station | 237.0 |
| Johnston Station | 261.0 |
| Davisboro | 302.0 |
| Sunhill Station | 362.0 |
| Tennille Station | 477.0 |
| Oconee Station | 228.0 |
| Toombsboro | 237.0 |
| Oconee River | 221.0 |
| McIntyre | 264.0 |
| Gordon | 354.0 |
| Pulaski | 374.0 |
| Griswold | 476.0 |
| Macon | 310.0 |
| River Flat | 300.0 |
| Point, 163 Mile-post | 300.0 |
| Top of Ridge | 481.0 |
| Summit | 475.0 |
| Passenger Depot, Macon | 377.0 |
| Switch-back, M. & W. | 401.0 |
| Holt Place | 584.0 |
| Howards | 485.0 |
| Mims House | 598.0 |
| Crawford | 621.0 |
| Winn Road-crossing | 669.0 |
| Trammell's | 590.0 |

### CENTRAL OF GEORGIA R. R.
*(Continued)*

| Station | Elevation[1] |
|---|---|
| Mrs. Thomas's | 759.0 |
| Collier's Station | 781.0 |
| The Jossey Estate | 777.0 |
| Gardner | 857.0 |
| Goggins Station | 842.0 |
| Goodwins | 905.0 |
| Road-crossing, 232 Mile-post | 933.0 |
| Barnesville | 903.0 |
| Milner Station | 894.0 |
| Simms' Place | 881.0 |
| Gilbert Weaver's | 882.0 |
| I. Andrews' | 944.0 |
| B. F. Sorcircy | 979.0 |
| Cunningham | 997.0 |
| Thornton Station | 915.0 |
| Griffin | 1,004.0 |
| Cox Land | 1,000.0 |
| Pat Sullivan's | 920.0 |
| Ben. Barfield's | 975.0 |
| S. P. Campbell | 937.0 |
| G. Dorsey's | 1,012.0 |
| Love Ivy Station | 1,002.0 |
| J. McVickers | 937.0 |
| Jonesboro | 995.0 |
| Atlanta | 1,085.0 |

### EDEN EXTENSION, CENTRAL R. R.

| Station | Elevation[1] |
|---|---|
| Meldrim Station | 39.3 |
| Black Creek | 14.3 |
| Ogeechee River | 14.3 |
| Ogeechee River, East Bank | 30.3 |
| Ogeechee River, West Bank | 29.3 |
| Cuyler | 37.3 |

[1] In feet.

EDEN EXTENSION, CENTRAL R. R.

(*Continued*)

| Station | Elevation [1] |
|---|---|
| East Bank, Black Creek | 45.3 |
| West Bank, Black Creek | 59.3 |
| Road-crossing, 21¾ Miles | 76.3 |
| Section-house | 74.3 |
| Ellabell | 93.5 |
| Malden Branch | 58.5 |
| Savannah Road Crossing | 76.5 |
| Toney Branch, 26⅓ Miles | 63.5 |
| Toney Branch, 27 Mile-post | 69.5 |
| Main Run | 79.5 |
| Pembroke Station | 101.5 |
| Savage Creek | 96.5 |
| Sam Baconfield's | 110.5 |
| Gin Branch | 99.5 |
| John Baconfield's | 107.5 |
| Harvey Branch | 106.5 |
| Savannah Road Crossing | 114.5 |
| Dry Branch | 107.5 |
| Uphaupee Station | 162.5 |
| Cannouchee River | 63.5 |
| Conly Station | 184.5 |
| Mt. Vernon & Savannah Road Crossing, 45 Mile-post | 180.5 |
| Branch, 45½ Miles | 155.5 |
| Mt. Vernon and Savannah Road Crossing 48¾ | 194.5 |
| Mt. Vernon and Savannah Road Crossing, 49¼ Miles | 196.5 |
| Bull Creek Ch. Road Crossing | 194.5 |
| Haw Pond | 201.5 |
| Bellville Station | 186.5 |
| Branch, 54½ Miles | 206.5 |
| Manassas Station | 217.5 |
| Collins Station | 238.5 |
| Branch, 61¾ Miles | 196.5 |

EDEN EXTENSION, CENTRAL R. R.

(*Continued*)

| Station | Elevation [1] |
|---|---|
| Bracewell Creek, 62½ Miles | 168.5 |
| Bed of Bracewell Creek, 64 Mile-post | 184.5 |
| East Side of Valley | 128.5 |
| Ohoopee River | 99.5 |
| West Side of Valley | 115.5 |
| Ohoopee Station | 187.5 |
| Branch, 69½ Miles | 149.5 |
| Mill Branch, 76¼ Miles | 127.5 |
| Pendleton Creek | 110.5 |
| East Side of Valley | 140.5 |
| West Side of Valley | 138.5 |
| Branch, 72 Mile-post | 153.5 |
| Branch, 72¼ Miles | 160.5 |
| Branch, 72½ " | 160.5 |
| Lyons Station | 254.5 |
| McLeod's House | 253.5 |
| Branch, 81 Mile-post | 257.5 |
| Branch, 82¾ Miles | 249.5 |
| Branch, 83¼ " | 246.5 |
| Black Creek | 244.5 |
| Rocky Creek | 258.5 |

AUGUSTA DIVISION, CENTRAL R. R.

| Station | Elevation [1] |
|---|---|
| Millen | 157.5 |
| Buckhead Creek | 145.0 |
| Road-crossing, 82¾ Miles | 182.0 |
| Lawton | 225.6 |
| Hines' Mill Creek | 199.2 |
| Road-crossing, 84¾ Miles | 212.2 |
| Road-crossing, 88 Mile-post | 252.0 |
| Long Branch | 242.0 |
| Branch, 89¼ Miles | 255.0 |
| Ridge, 89¾ " | 275.0 |
| Public Road, 90¼ Miles | 263.4 |

[1] In feet.

HIGH FALLS OF THE TOWALIGA, MONROE COUNTY, GEORGIA.

# RAILROAD ELEVATIONS

| AUGUSTA DIVISION, CENTRAL R. R. (Continued) | | SOUTHWESTERN DIVISION, CENTRAL R. R. | |
|---|---|---|---|
| Station | Elevation [1] | Station | Elevation [1] |
| Lumpkin Station | 264.4 | Passenger Depot, Macon | 377.0 |
| Branch, 91 Mile-post | 252.0 | Starting Point | 328.0 |
| Carter's Branch | 253.0 | Tobesofkee Ridge | 382.2 |
| Proctor's Branch | 277.2 | Tobesofkee Creek | 313.0 |
| Ship Ridge | 283.5 | Ridge, 198 Mile-post | 396.9 |
| Pond's Branch | 277.9 | Walden Station | 390.6 |
| Thomas Station | 285.7 | Echeconnee Creek | 303.1 |
| Road-crossing, 96¼ Miles | 300.7 | Byron Station | 515.6 |
| "       "    97¼  " | 302.2 | Powersville | 406.3 |
| McIntosh Creek | 262.8 | Fort Valley | 531.3 |
| Waynesboro Station | 286.7 | Marshallville | 500.0 |
| Briar Creek | 199.7 | Winchester | 375.0 |
| Gouns Cut | 284.9 | Montezuma | 300.0 |
| McBean Creek | 140.9 | Flint River | 303.1 |
| McBean Station | 134.6 | Oglethorpe | 313.0 |
| Dickerson Canal | 127.6 | Ridge, 249 Mile-post | 398.0 |
| Little McBean Creek | 117.2 | Sweet Water Creek | 366.0 |
| McBean Mill | 126.6 | Americus Ridge | 469.0 |
| Barney Bluff | 124.2 | Americus | 350.0 |
| Valley, 119¼ Miles | 122.1 | Smithville Ridge | 372.0 |
| Ridge, 120¼ | 140.9 | Smithville | 319.0 |
| Road-crossing, 121 Mile-post | 133.6 | Albany | 184.4 |
| Spring Creek | 119.8 | East Albany | 186.0 |
| Allen's Station | 139.2 | | |
| Butter Creek | 141.5 | | |

[1] In feet.

NOTE. — It is impossible to harmonize the data of all railroads, centering in Macon; because the points, whose elevations are given, cannot be definitely located and united, by a line of levels. These elevations have been tied, when possible, in regions of level ground, rather than in the hills of Middle Georgia, where a slight error in location would make a discrepancy of several feet in elevation. Waycross, Valdosta, Tifton, Albany, Smithville and Thomasville have been chosen, for the tie-points; but harmony, at the above named places, causes discrepancies at Macon and Atlanta, that can be explained, only on the theory of gross errors in working out the levels in the original surveys.

### EUFAULA BRANCH, SOUTHWESTERN DIVISION, CENTRAL. R. R.
*(Continued)*

| Station | Elevation [2] |
|---|---|
| Smithville | 319.0 |
| Kinchafoonee Creek | 265.0 |
| East Chickasawhachee Creek | 334.0 |
| Middle Prong of Chickasawhachee Creek | 334.0 |
| West Prong of Chickasawhachee Creek | 312.0 |
| 100 Mile-post | 362.0 |
| Station, 292 Mile-post | 326.0 |
| Creek, 295½ Miles | 283.0 |
| Station, 298 Mile-post | 379.0 |
| Double Branch | 387.0 |
| Pachitla Creek | 340.0 |
| Cuthbert Depot | 432.0 |
| Railroad Junction | 469.0 |
| 125 Mile-post | 274.0 |
| Station, 319½ Miles | 235.0 |
| Stream, 321 Mile-post | 212.0 |
| Station, 324 Mile-post | 289.0 |
| Tobenannee Creek | 214.0 |
| Georgetown Depot | 189.0 |
| Near River, 332½ Miles | 178.0 |
| Beyond River, 333 Mile-post | 199.0 |
| Eufaula, Alabama | 211.0 |

### FORT GAINES BRANCH, SOUTHWESTERN DIVISION, CENTRAL. R. R.[1]

| Station | Elevation [2] |
|---|---|
| Junction, 311 Mile-post | 469.0 |
| 126 Mile-post | 424.0 |
| Samocheehabbee Creek | 161.0 |
| Fort Gaines | 252.0 |

### MUSCOGEE R. R., SOUTHWESTERN DIVISION, CENTRAL. R. R.

| Station | Elevation [2] |
|---|---|
| Fort Valley | 531.0 |
| Flint River | 337.0 |
| Reynold's | 433.0 |
| 52 Mile-post | 506.0 |
| Butler Station | 650.0 |
| Station, 250 Mile-post | 666.0 |
| Bostwick | 669.0 |
| Geneva | 600.0 |
| Upatoie | 432.0 |
| Upatoie Creek | 413.0 |
| Keaton | 382.0 |
| Station, 267 Mile-post | 382.0 |
| Far River | 382.0 |
| Kendall's Mill | 392.0 |
| Cox Creek | 397.0 |
| Station, 273 Mile-post | 460.0 |
| Randall Creek | 313.0 |
| Station, 276 Mile-post | 460.0 |
| Dozier Creek | 439.0 |
| Bull Creek | 378.0 |
| Station, 281 Mile-post | 322.0 |
| Columbus | 262.0 |

### MACON & DUBLIN R. R.

| Station | Elevation [2] |
|---|---|
| 2 Mile-post, Macon & North. R. R. | 516.0 |
| Swift Creek | 536.0 |
| Branch, 5 Mile-post | 538.0 |
| Bottom of Swift Creek | 512.0 |
| Cut, Crosstie, 5¼ Miles | 545.0 |
| " Ground Surface | 575.0 |
| Bottom of Branch, 7 Mile-post | 539.0 |
| " " " 8¼ Miles | 570.0 |

[1] Datum :— Reduced to Fort Pulaski, Mean Low Tide, by adding constant 86 to all elevations.

[2] In feet.

## MACON & DUBLIN R. R.
*(Continued)*

| Station | Elevation[1] |
|---|---|
| Branch, 9 Mile-post | 575.0 |
| Branch Bottom | 564.0 |
| Dry Branch Station | 589.0 |
| Branch Bottom | 659.0 |
| 1st Large Cut, 12 Mile-post | 723.0 |
| Ground-Surface | 769.0 |
| 2nd Large Cut, 12½ Miles | 752.0 |
| Ground Surface | 793.0 |
| Pike's Peak Station | 755.0 |
| Branch Bottom, 12¾ Miles | 713.0 |
| Fitzpatrick Station | 762.0 |
| Branch, Ground Surface, 17¼ Miles | 738.0 |
| Branch, 18¼ Miles | 767.0 |
| Branch Bottom | 751.0 |
| Macon Road Crossing | 745.0 |
| Allentown Road Crossing | 752.0 |
| Jeffersonville Station | 747.0 |
| Road-crossing | 734.0 |
| 24 Mile-post | 710.0 |
| Branch, 26½ Miles | 634.0 |
| Palmetto Creek Bottom | 591.0 |
| Gallimore Station | 594.0 |
| Turkey Creek, 29 Mile-post | 575.0 |
| Hughes Station | 572.0 |
| Allentown Station | 651.0 |
| Montrose Station | 612.0 |
| Elsie Station | 546.0 |
| Branch, 44 Mile-post | 516.0 |
| Turkey Creek, 46¼ Miles | 439.0 |
| Spring Branch Bottom | 424.0 |
| Moore Station | 479.0 |
| Dublin | 452.0 |
| Moore Street, Dublin | 442.0 |
| Lawrence Street, " | 442.0 |
| Jefferson Street, " | 440.0 |

## MACON & DUBLIN R. R.
*(Continued)*

| Station | Elevation[1] |
|---|---|
| Oconee River Bluff | 413.0 |
| High-water Mark | 400.0 |
| West Bank of Oconee | 394.0 |
| Bottom of Oconee | 362.0 |
| East Bank of Oconee | 396.0 |
| 12 Mile-post | 772.0 |
| 13 " " | 773.0 |
| 14 " " | 772.0 |
| 15 " " | 764.0 |
| 16 " " | 783.0 |
| 17 " " | 761.0 |
| 18 " " | 765.0 |
| 19 " " | 749.0 |
| 20 " " | 764.0 |
| 21 " " | 751.0 |
| 22 " " | 750.0 |
| 23 " " | 732.0 |
| 24 " " | 710.0 |
| 25 " " | 662.0 |
| 26 " " | 598.0 |
| 27 " " | 586.0 |
| 28 " " | 575.0 |
| 29 " " | 632.0 |
| 30 " " | 664.0 |
| 31 " " | 658.0 |
| 32 " " | 643.0 |
| 33 " " | 632.0 |
| 34 " " | 632.0 |
| 35 " " | 602.0 |
| 36 " " | 608.0 |
| Ravine, 55 Mile-post | 409.0 |
| Shaddock Creek | 404.0 |
| Mt. Vernon Road | 408.0 |
| Pugh's Creek Bottom | 404.0 |
| Branch, 67¼ Miles | 445.0 |

[1] In feet.

| MACON & DUBLIN R. R. (*Continued*) | | MACON & DUBLIN R. R. (*Continued*) | |
|---|---|---|---|
| Station | Elevation[1] | Station | Elevation[1] |
| Branch, 68½ Miles | 509.0 | Ridge, 98½ Miles | 428.0 |
| Blackville Road | 512.0 | Branch, 98¾ " | 400.0 |
| Alligator Creek | 500.0 | Ridge, 99¼ " | 421.0 |
| Branch, 72¼ Miles | 495.0 | Branch, 99½ Miles | 400.0 |
| " 73 Mile-post | 484.0 | Ridge, 100½ " | 465.0 |
| Road, 74½ Miles | 460.0 | Road, " " | 460.0 |
| Branch, 75¼ " | 453.0 | Wolf Creek | 484.0 |
| Road, 77¼ " | 457.0 | 1st Ridge, 101½ Miles | 453.0 |
| Pendleton Creek | 440.0 | 2nd " " " | 453.0 |
| Branch, 78 Mile-post | 442.0 | 1st Branch of Wolf Creek | 429.0 |
| Ridge, 80 Mile-post | 489.0 | Branch, 103½ Miles | 395.0 |
| Branch, 80 Mile-post | 452.0 | Ridge Road, 105¼ Miles | 419.0 |
| Red Bluff Creek | 420.0 | Branch, 105½ Miles | 397.0 |
| Ridge, 82 Mile-post | 474.0 | Road, 106 Mile-post | 398.0 |
| Branch, 82¼ Miles | 449.0 | Cannouchee River | 344.0 |
| Branch, 83 Mile-post | 441.0 | High-water Mark | 348.0 |
| " 84½ Miles | 461.0 | Branch, 110¾ Miles | 354.0 |
| " 86 Mile-post | 477.0 | Reidsville Road | 360.0 |
| " 88¼ Miles | 411.0 | 10-mile Creek | 336.0 |
| " 89½ Miles | 377.0 | Road, 115¼ Miles | 376.0 |
| Low-grounds | 366.0 | " 123¾ " | 364.0 |
| Bottom of Ohoopee | 354.0 | Lot's Creek | 305.0 |
| High-water Mark | 372.0 | Road, 129 Mile-post | 350.0 |
| Ridge, 94 Mile-post | 440.0 | Bullock's Bay | 328.0 |
| Jack's Creek | 356.0 | Bay Gall | 310.0 |
| Branch, 97¼ Miles | 371.0 | Road, 133¼ Miles | 320.0 |
| " 97¾ " | 388.0 | Road, 134¼ " | 319.0 |

[1] In feet.

## ELEVATIONS

The following are the elevations above the average sea-level of some of the prominent mountains and other points of interest in the State, determined by the United States Coast and Geodetic Survey:—

| | Elevation in feet |
|---|---|
| Sitting Bull (middle summit of Nantahala, Towns county) | 5,046 |
| Mona (east summit of Nantahala, Towns county) | 5,039 |
| Enota, in Towns county | 4,797 |
| Rabun Bald, in Rabun county | 4,718 |
| Blood, in Union county | 4,468 |
| Tray, in Habersham county | 4,403 |
| Cohutta, in Fannin county | 4,155 |
| Dome, in Towns county | 4,042 |
| Grassy, in Pickens county | 3,290 |
| Tallulah (northwest summit), in Habersham county | 3,172 |
| Tallulah (southeast summit), in Habersham county | 2,849 |
| Yonah, in White county | 3,167 |
| Walker, in Lumpkin county | 2,614 |
| Lookout (at High Point), in Walker county | 2,390 |
| Pine Log, in Bartow county | 2,346 |
| Lookout (at Round Mountain), in Walker county | 2,331 |
| Pigeon (at High Point), in Walker county | 2,329 |
| Skit | 2,075 |
| Sawnee, in Forsyth county | 1,968 |
| Kennesaw, in Cobb county | 1,809 |
| Stone Mountain, in DeKalb county | 1,686 |
| Sweat | 1,693 |
| Lavender, in Floyd county | 1,680 |
| Cleveland Church, in White county | 1,616 |
| Taylor's Ridge, in Chattooga county | 1,556 |
| Dahlonega Agricultural College | 1,518 |
| Mt. Alto, in Floyd | 1,505 |
| Clarkesville Court House, in Habersham county | 1,478 |
| Carnes Mountain, in Polk county | 1,296 |

# APPENDIX

## INTRODUCTION

### By W. S. YEATES, State Geologist

Since the work of compiling the report on the *Water-powers of Georgia*, which forms the first part of this bulletin, was completed, a great deal of hydrographic work has been done in Georgia, by the co-operation of this Survey with the U. S. Geological Survey, mentioned in the letter of transmittal.[1]  As indicated, in this letter, it was the intention of the State Geologist, to use the results of that work, in a second bulletin, to be published, as soon as sufficient field-data had been collected.  As it has taken a much longer time to bring out this bulletin, than was at first anticipated, it is best to include, in the form of an appendix to the first report, the work since accomplished in the field, bringing the subject up to date.[2]

In the fall of 1895, Mr. B. M. Hall, who had been employed, by this Survey, as Special Assistant, to compile the report on the Water-powers of Georgia, embraced in the first part of this bulletin, was appointed Hydrographer for the U. S. Geological Survey, in charge of the work on the rivers of Georgia, Florida, Alabama and Tennessee, under the direction of Mr. F. H. Newell, Chief of the Hydrographic Division of the U. S. Geological Survey.  Subsequently, the plan of co-operation, referred to, was agreed upon; and all work, done by the two Surveys, since Mr. Hall began, in the latter part of 1895, is here presented.

---

[1] See page 5.        [2] July 1st, 1897.

During this time, Mr. Hall has been regularly assisted, in the field-work, by Messrs. Max Hall, Olin P. Hall and P. A. Dallis; while the following river-observers have been employed, at the various stations, indicated:—

| Observer | Station | River |
|---|---|---|
| Col. S. M. Carter | Carter's | Coosawattee |
| J. H. Lowry | Oakdale | Chattahoochee |
| C. E. Melton | West Point | " |
| J. P. Mercer | Macon | Ocmulgee |
| S. M. Barnett | Resaca | Oostanaula |
| Peter Pfeiffer | S. A. L. Bridge | Savannah |
| J. A. Low | Canton | Etowah |
| J. L. Cary | Carey | Oconee |
| U. S. Weather Bureau | Dublin | " |
| J. A. Moore | Molena | Flint |

These gentlemen have been paid a small amount, as compensation for their services, except Col. Carter, who kindly consented, to act as observer at Carter's Station, without compensation; and they have made weekly reports, on daily observations, both to the U. S. Geological Survey and to the Geological Survey of Georgia. By courtesy of the U. S. Weather Bureau, observations at Dublin have been furnished, without cost to either Survey; but, as this station has been discontinued by the Weather Bureau, further observations, here, will require the employment of an observer.

The plan of co-operation has resulted in accomplishing a much greater work, for both Surveys; and it is proposed, to continue this plan, for collecting data, for our next bulletin, on this subject. It is the very liberal policy of the Director of the U. S. Geological Survey towards the State Surveys, that has made it possible, for the Geological Survey of Georgia to collect so much data, at so small an expense to the State; and further co-operation, along other lines of work, will probably be effected, in the near future.

## METHODS AND RESULTS OF RECENT WORK

### By B. M. HALL, Hydrographer

The following is a brief statement of the methods, adopted, and the results accomplished, in the field-work, done, since the foregoing report on the Water-powers of Georgia was compiled : —

This appendix deals, exclusively, with the amount of water, flowing in the streams, and gives a safe basis, for calculation of low-water volumes, at the separate water-powers, described in the foregoing report; the same being applicable, only to the driest years, ever known in this region. The work was begun, in the Autumn of 1895; and it has continued, without ceasing, to the present time. Its object has been to obtain a knowledge of the exact amount of water, flowing in the streams, at all seasons of the year, in order to arrive at their value for water-power, irrigation, mining, municipal supply etc. Certain convenient stations have been established on the important rivers. At each of these stations, a gauge-rod is set, to show the fluctuations of the stream; and a gauge-reader is employed, to observe the height of water on the gauge, every morning, at the same hour, and to make a weekly report of the same to the Hydrographer-in-charge. From time to time, the Hydrographer, or one of his field-assistants, visits the station, and makes an accurate discharge-measurement of the stream, noting the height of the water on the gauge, at the time the discharge-measurement is made. After a large number of discharge-measurements have been made, at different gauge-heights, a rating-table is made, from the data thus obtained, which gives the amount of water, flowing in the stream, at that station, for any gauge-height, shown on the rod. Thus, by inspection of the table of daily gauge-heights, the flow of the stream is shown, for every day in the year, or years, covered by observations of gauge-height. As the main object of the work, so far, has been to get the value of the streams, for water-power, special attention has been given to low-water measurements; and the rating-tables do not cover the highest stages of water.

In making discharge-measurements, the velocities are taken, at all points of the section, with the latest improved electric current-meters; and accurate cross-sections are made, from soundings, 10 feet apart.

The minimum low-water measurements, given here, were made in the Autumn of 1896, when the streams were at the lowest stage, that they have reached, for many years — a minimum stage, that they probably reach, only once or twice in a century. This will be shown by a study of the Atlanta rainfall table, from July 1870 to December 1896, inclusive, published on page 18 of this bulletin, giving, for 26 years, an average annual rainfall of 50.96 inches. It gives positive evidence, that the streams of this region were lower, during the year 1896, than at any time since 1870. There has been a continuous accumulating deficiency since 1890, which, however, did not begin to make a visible impression on the streams, until 1893, though it naturally affected the supply of ground-water, available for the following years. But, on top of this deficiency, has come a period of four years, from 1893 to 1896, inclusive, in which the average annual rainfall was 39.35 inches, distributed as follows: — Spring — 9.87; Summer — 12.43; Autumn — 6.24; and Winter — 10.81. This distribution shows, that the greatest rainfall, during the period, named, has been in Summer, when the amount of water, taken up by vegetation and evaporation, was greatest. The fact, that these conditions have produced good crops, would naturally prevent most people from recognizing the years, named, as exceptionally dry ones; but it is stated, by the oldest inhabitants, that the streams and wells were lower, in the Autumn of 1896, than they have ever seen them, since the year 1845. It must, therefore, be expected, that the minimum discharges, given below, will be much smaller, than those found by Mr. C. C. Anderson, late Assistant State Geologist, in 1891 and 1892, when the streams were at their average stage.

# THE SAVANNAH BASIN

## SAVANNAH RIVER

### Seaboard Air Line R. R. Bridge Station, Elbert County Georgia

On August 4th, 1896, a regular station was established on The Savannah River, in Elbert county, Georgia, at the Seaboard Air Line R. R. Bridge. The drainage area, or water-shed, above this point, is 2,695 square miles. Mr. Peter Pfeiffer of Calhoun Falls, S. C., the nearest railroad station, was made observer. The following represents the work done at this station :—

### DISCHARGE MEASUREMENTS

| No. | Date | Measurement Made by | Meter Number | Gauge-height in Feet | Area of Section in Square Feet | Mean Velocity in Feet per Second | Discharge in Cubic Feet per Second |
|---|---|---|---|---|---|---|---|
| 1 | 1896 Aug. 4 | Max Hall | 16 | 2.40 | 2,278 | 1.170 | 2,665 |
| 2 | Sept. 22 | " " | 11 | 1.77 | 1,488 | 0.980 | 1,531 |
| 3 | Oct. 31 | " " | 11 | 2.10 | 1,889 | 1.090 | 2,054 |
| 4 | 1897 Jan. 20 | B. M. Hall | 8 | 2.90 | 2,173 | 1.935 | 4,204 |
| 5 | Apr. 28 | Max Hall | 91 | 3.21 | 2,811 | 2.290 | 6,446 |
| 6 | June 12 | " " | 11 | 2.80 | 2,606 | 1.714 | 4,469 |

## DAILY GAUGE-HEIGHT[1]

### Peter J. Pfeiffer, *Observer*

| | 1896 | | | | | 1897 | | | | | |
|---|---|---|---|---|---|---|---|---|---|---|---|
| | Aug. | Sept. | Oct. | Nov. | Dec. | Jan. | Feb. | Mar. | Apr. | May | June |
| 1 |  | 2.00 | 2.00 | 2.20 | 5.60 | 2.50 | 2.80 | 3.00 | 5.40 | 3.80 | 2.20 |
| 2 |  | 1.90 | 1.95 | 2.15 | 5.00 | 2.40 | 5.20 | 2.95 | 6.90 | 5.65 | 2.15 |
| 3 |  | 1.85 | 1.90 | 2.05 | 4.95 | 2.40 | 4.00 | 2.80 | 5.20 | 4.30 | 3.05 |
| 4 | 2.40 | 1.80 | 1.95 | 3.00 | 5.15 | 2.40 | 3.60 | 2.80 | 4.75 | 3.95 | 4.40 |
| 5 | 2.30 | 1.95 | 1.90 | 5.65 | 5.00 | 2.35 | 3.25 | 2.75 | 11.60 | 3.85 | 4.10 |
| 6 | 2.15 | 3.85 | 1.85 | 7.15 | 4.05 | 2.35 | 6.00 | 2.65 | 13.35 | 3.80 | 3.10 |
| 7 | 2.10 | 3.00 | 1.90 | 4.75 | 3.50 | 2.30 | 8.55 | 6.80 | 8.15 | 3.70 | 2.95 |
| 8 | 2.00 | 2.90 | 1.75 | 3.00 | 3.75 | 2.25 | 7.20 | 4.65 | 4.95 | 3.65 | 3.05 |
| 9 | 2.05 | 2.40 | 1.70 | 2.60 | 3.65 | 2.25 | 5.05 | 4.20 | 4.05 | 3.40 | 3.25 |
| 10 | 2.10 | 2.25 | 1.70 | 2.45 | 3.20 | 2.25 | 4.10 | 4.00 | 4.00 | 3.29 | 3.05 |
| 11 | 2.05 | 2.20 | 2.00 | 2.30 | 2.85 | 2.20 | 3.85 | 4.40 | 4.15 | 3.15 | 2.95 |
| 12 | 2.00 | 2.30 | 2.20 | 2.20 | 2.60 | 2.20 | 5.15 | 5.50 | 4.10 | 3.08 | 2.80 |
| 13 | 1.95 | 2.15 | 2.50 | 5.60 | 2.45 | 2.30 | 4.40 | 7.75 | 4.00 | 3.00 | 3.00 |
| 14 | 2.80 | 2.10 | 2.40 | 4.10 | 2.55 | 3.05 | 4.10 | 7.25 | 3.95 | 3.05 | 2.85 |
| 15 | 3.10 | 2.05 | 2.15 | 3.60 | 3.85 | 2.75 | 4.00 | 6.00 | 3.95 | 3.10 | 2.80 |
| 16 | 2.30 | 2.00 | 2 00 | 3.25 | 3.20 | 3.60 | 4.05 | 5.20 | 3.85 | 3.15 | 3.00 |
| 17 | 2.10 | 2.00 | 1.95 | 3.00 | 3.40 | 2.55 | 3.95 | 4.15 | 3.80 | 3 10 | 3.05 |
| 18 | 2.05 | 1.95 | 1.90 | 2.90 | 3.10 | 3.35 | 3.80 | 3.85 | 3.75 | 3 05 | 2.95 |
| 19 | 2.05 | 1.90 | 1.85 | 2.65 | 3.00 | 3.10 | 3.65 | 3.50 | 3.65 | 3.00 | 2 85 |
| 20 | 2.00 | 1.85 | 1.80 | 2.40 | 2.85 | 2.90 | 3.70 | 4.00 | 3.50 | 2.95 | 2.80 |
| 21 | 1.90 | 1.80 | 1.80 | 2.25 | 2.80 | 5.40 | 3 50 | 5.35 | 3.40 | 2.90 | 2.70 |
| 22 | 1.85 | 1.75 | 1.75 | 2.35 | 2.75 | 3.95 | 3.35 | 4.40 | 3.35 | 2 90 | 2.65 |
| 23 | 1.80 | 2.50 | 1.75 | 2.30 | 2.65 | 3.60 | 4.05 | 4.10 | 3.30 | 2.85 | 2.55 |
| 24 | 1.75 | 2.40 | 2.15 | 2.30 | 2.60 | 3.20 | 3.80 | 4.00 | 3.25 | 2.75 | 2.55 |
| 25 | 1.75 | 2.35 | 2.05 | 2.30 | 2.55 | 3.10 | 4.00 | 3.90 | 3.25 | 2.70 | 2.45 |
| 26 | 2.00 | 2.25 | 2.00 | 2.25 | 2.50 | 3.00 | 3.90 | 3.65 | 3.30 | 2.60 | 2.50 |
| 27 | 2.45 | 2.00 | 1.95 | 2.25 | 2.45 | 2.95 | 3.45 | 3.05 | 3.25 | 2.55 | 2.40 |
| 28 | 2.00 | 1.90 | 1.90 | 3 20 | 2.40 | 2.95 | 2.20 | 3.40 | 3.20 | 2.40 | 2.30 |
| 29 | 1.95 | 1.95 | 1.85 | 2.30 | 2.40 | 2.90 |  | 3.25 | 3.25 | 2.35 | 3.50 |
| 30 | 1.90 | 2.00 | 1.95 | 2.95 | 2.40 | 2.90 |  | 3.25 | 3.40 | 2.25 | 2.95 |
| 31 | 1.85 |  | 2.10 |  | 2.35 | 2.75 |  | 3.30 |  | 2.20 |  |

[1] In feet.

## APPENDIX

### RATING-TABLE

*Drainage Area, 2,695 Square Miles*

| Gauge-height in Feet | Discharge in Cubic Feet per Second | Gauge-height in Feet | Discharge in Cubic Feet per Second | Gauge-height in Feet | Discharge in Cubic Feet per Second | Gauge-height in Feet | Discharge in Cubic Feet per Second |
|---|---|---|---|---|---|---|---|
| 1.70 | 1,480 | 2.20 | 2,260 | 2.70 | 3,700 | 3.20 | 6,350 |
| 1.80 | 1,580 | 2.30 | 2,470 | 2.80 | 4,230 | 3.30 | 6,880 |
| 1.90 | 1,700 | 2.40 | 2,690 | 2.90 | 4,760 | 3.40 | 7,410 |
| 2.00 | 1,850 | 2.50 | 2,930 | 3.00 | 5,290 | 3.50 | 7,940 |
| 2.10 | 2,050 | 2.60 | 3,230 | 3.10 | 5,820 | | |

The minimum discharge per square mile of drainage area is 0.55 cubic feet per second.

### AUGUSTA, GEORGIA

The only other discharge measurement, made on the Savannah River, so far, was at *Augusta, Georgia*, at the North Augusta bridge.

### DISCHARGE MEASUREMENT

| No. | Date | Measurement Made by | Meter Number | Gauge-height in Feet[1] | Area of Section in Square Feet | Mean Velocity in Feet per Second | Discharge in Cubic Feet per Second |
|---|---|---|---|---|---|---|---|
| 1 | 1896 Oct. 3 | B. M. Hall | 8 | 5.42 | 3,178 | 0.992 | 3,154 |

[1] On Augusta city-gauge.

*APPENDIX*

# BROAD RIVER

### CARLTON STATION, MADISON COUNTY, GEORGIA

This station, at the Seaboard Air Line bridge, over the North Broad River, was established, May 27th, 1897; and discharge measurements were then begun; but the gauge-observer Mr. S. P. Power, Jr., does not begin his regular duties, until July 1st. The measurements made, so far, are:—

#### DISCHARGE MEASUREMENTS

| No | Date | Measurement Made by | Meter Number | Gauge-height in Feet | Area of Section in Square Feet | Mean Velocity in Feet per Second | Discharge in Cubic Feet per Second |
|---|---|---|---|---|---|---|---|
| 1 | 1897 May 27 | Max Hall | 91 | 2.10 | 594 | 1.004 | 596 |
| 2 | June 22 | " " | 91 | 1.92 | 604 | 0.960 | 580 |

The great number of fine water-powers in the Savannah Basin are accessible by the Southern Railway and the Seaboard Air Line and the Georgia Railroads.

# THE ALTAMAHA BASIN

## OCONEE RIVER

### Cary Station, Greene County, Georgia

This station was established, October 29th, 1896, at the Georgia Railroad bridge across the Oconee River, just below the mouth of the Apalachee River. The drainage area above this point is 1,346 square miles. This station is about 30 miles above Milledgeville. With Mr. J. L. Cary, as gauge-observer, the following work has been done at the station: —

#### DISCHARGE MEASUREMENTS

*Drainage Area, 1,346 Square Miles*

| No. | Date | Measurement Made by | Meter Number | Gauge-height in Feet | Area of Section in Square Feet | Mean Velocity in Feet per Second | Discharge in Cubic Feet per Second |
|---|---|---|---|---|---|---|---|
|  | 1896 |  |  |  |  |  |  |
| 1 | Oct. 29 | Max Hall | 11 | 1.68 | 735 | 0.880 | 644 |
| 2 | Nov. 17 | B. M. Hall | 8 | 2.08 | 702 | 1.190 | 836 |
| 3 | Nov. 25 | " " | 8 | 1.90 | 715 | 1.110 | 795 |
|  | 1897 |  |  |  |  |  |  |
| 4 | Jan. 18 | B. M. Hall | 8 | 4.95 | 1,344 | 2.468 | 3,318 |
| 5 | Mar. 18 | " " | 91 | 5.15 | 1,417 | 3.000 | 4,257 |
| 6 | Apr. 29 | Max Hall | 91 | 2.40 | 963 | 2.070 | 1,992 |
| 7 | May 28 | B. M. Hall | 14 | 2.10 | 701 | 1.494 | 1,047 |
| 8 | June 9 | Max Hall | 11 | 2.50 | 949 | 1.986 | 1,885 |

## APPENDIX

### DAILY GAUGE-HEIGHT[1]

J. L. Cary, *Observer*

|    | 1896 |      |      | 1897 |      |      |       |      |      |      |
|----|------|------|------|------|------|------|-------|------|------|------|
|    | Oct. | Nov. | Dec. | Jan. | Feb. | Mar. | Apr.  | May  | June | July |
| 1  | . .  | 2.10 | 5.10 | 2.10 | 2.50 | 3.20 | 3.30  | 4.00 | 2.10 | . .  |
| 2  | . .  | 1.90 | 4.80 | 1.90 | 3.80 | 3.20 | 4.90  | 3.80 | 2.10 | . .  |
| 3  | . .  | 1.70 | 4.40 | 2.10 | 3.70 | 3.10 | 6.10  | 3.30 | 2.10 | . .  |
| 4  | . .  | 2.10 | 4.20 | 2.00 | 3.60 | 3.90 | 5.60  | 2.80 | 2.20 | . .  |
| 5  | . .  | 2.70 | 3.70 | 2.10 | 3.30 | 3.00 | 8.80  | 2.60 | 2.20 | . .  |
| 6  | . .  | 2.30 | 3.40 | 2.00 | 4.60 | 2.90 | 14.40 | 2.50 | 2.30 | . .  |
| 7  | . .  | 2.20 | 3.20 | 2.10 | 5.00 | 6.40 | 12.40 | 2.30 | 2.20 | . .  |
| 8  | . .  | 1.80 | 3.00 | 2.10 | 4.60 | 7.80 | 7.30  | 2.30 | 2.20 | . .  |
| 9  | . .  | 1.80 | 2.90 | 1.80 | 3.80 | 6.80 | 5.40  | 2.30 | 2.50 | . .  |
| 10 | . .  | 1.80 | 2.70 | 1.90 | 3.30 | 4.40 | 5.50  | 2.30 | 2.30 | . .  |
| 11 | . .  | 1.80 | 2.60 | 2.00 | 3.10 | 4.00 | 4.50  | 2.20 | 2.20 | . .  |
| 12 | . .  | 1.90 | 2.50 | 1.90 | 5.90 | 4.20 | 4.00  | 2.30 | 2.00 | . .  |
| 13 | . .  | 1.80 | 2.40 | 2.00 | 6.60 | 7.70 | 3.50  | 2.30 | 1.80 | . .  |
| 14 | . .  | 1.90 | 2.30 | 2.70 | 5.30 | 10.40| 3.50  | 2.40 | 1.70 | . .  |
| 15 | . .  | 1.90 | 4.00 | 4.30 | 4.40 | 12.20| 3.30  | 2.30 | 1.70 | . .  |
| 16 | . .  | 2.00 | 2.80 | 4.20 | 4.00 | 11.60| 3.30  | 2.30 | 1.60 | . .  |
| 17 | . .  | 2.08 | 2.40 | 3.40 | 4.20 | 8.60 | 3.30  | 2.30 | 1.60 | . .  |
| 18 | . .  | 2.00 | 2.20 | 4.50 | 3.60 | 5.50 | 3.00  | 2.20 | 1.60 | . .  |
| 19 | . .  | 2.00 | 2.40 | 4.80 | 3.30 | 4.20 | 2.90  | 2.20 | 1.80 | . .  |
| 20 | . .  | 1.90 | 2.30 | 4.00 | 3.00 | 5.30 | 2.80  | 2.20 | 2.40 | . .  |
| 21 | . .  | 1.80 | 2.20 | 6.00 | 3.80 | 5.50 | 2.70  | 2.00 | 2.10 | . .  |
| 22 | . .  | 1.90 | 2.20 | 7.80 | 4.00 | 4.60 | 2.70  | 1.90 | 2.00 | . .  |
| 23 | . .  | 2.00 | 2.10 | 6.80 | 3.80 | 4.60 | 2.60  | 2.00 | 1.70 | . .  |
| 24 | . .  | 1.95 | 2.00 | 4.30 | 4.70 | 4.70 | 2.60  | 2.20 | 1.50 | . .  |
| 25 | . .  | 1.90 | 2.00 | 3.30 | 5.30 | 4.20 | 2.70  | 2.10 | 2.20 | . .  |
| 26 | . .  | 1.90 | 2.10 | 3.10 | 5.20 | 3.70 | 2.50  | 2.10 | 2.00 | . .  |
| 27 | . .  | 1.80 | 2.00 | 2.80 | 4.20 | 3.20 | 2.50  | 2.10 | 1.80 | . .  |
| 28 | . .  | 1.90 | 1.90 | 2.80 | 3.50 | 3.20 | 2.50  | 2.00 | 1.60 | . .  |
| 29 | 1.68 | 2.90 | 2.10 | 2.60 | . .  | 3.00 | 2.50  | 2.10 | 1.50 | . .  |
| 30 | 1.68 | 3.90 | 2.00 | 2.40 | . .  | 3.00 | 2.90  | 2.10 | 1.60 | . .  |
| 31 | 1.68 | . .  | 2.00 | 2.50 | . .  | 3.40 | . .   | 2.20 | . .  | . .  |

[1] In feet.

## RATING-TABLE

*Drainage Area, 1,346 Square Miles*

| Gauge-height in Feet | Discharge in Cubic Feet per Second | Gauge-height in Feet | Discharge in Cubic Feet per Second | Gauge-height in Feet | Discharge in Cubic Feet per Second | Gauge-height in Feet | Discharge in Cubic Feet per Second |
|---|---|---|---|---|---|---|---|
| 1.70 | 650   | 2.70 | 2,010 | 3.70 | 2,730 | 4.70 | 3,328 |
| 1.80 | 720   | 2.80 | 2,100 | 3.80 | 2,800 | 4.80 | 3,420 |
| 1.90 | 800   | 2.90 | 2,200 | 3.90 | 2,840 | 4.90 | 3,550 |
| 2.00 | 900   | 3.00 | 2,280 | 4.00 | 2,880 | 5.00 | 3,720 |
| 2.10 | 1,000 | 3.10 | 2,360 | 4.10 | 2,940 | 5.10 | 3,910 |
| 2.20 | 1,100 | 3.20 | 2,440 | 4.20 | 3,000 | 5.20 | 4,150 |
| 2.30 | 1,280 | 3.30 | 2,510 | 4.30 | 3,050 | 5.30 | 4,350 |
| 2.40 | 1,480 | 3.40 | 2,560 | 4.40 | 3,100 |      |       |
| 2.50 | 1,560 | 3.50 | 2,620 | 4.50 | 3,170 |      |       |
| 2.60 | 1,850 | 3.60 | 2,680 | 4.60 | 3,230 |      |       |

The irregularity in this rating-table is caused by obstructions in the river, at the station, and by a mill-dam, about five miles below. For minimum discharge of river, see measurements at Milledgeville.

### DUBLIN STATION, LAURENS COUNTY, GEORGIA

This station is located at Dublin, Ga., at the Iron Highway bridge, belonging to Laurens county. Discharge measurements were begun May 5th, 1897. The following is a statement of the work done to date: —

## DISCHARGE MEASUREMENTS

| No. | Date | Measurement Made by | Meter Number | Gauge-height in Feet | Area of Section in Square Feet | Mean Velocity in Feet per Second | Discharge in Cubic Feet per Second |
|---|---|---|---|---|---|---|---|
| 1 | 1897 May 5 | B. M. Hall | 91 | 6.10 | 2,251 | 2.843 | 6,400 |
| 2 | June 7 | P. A. Dallis | 14 | 1.90 | 1,151 | 2.485 | 2,861 |
| 3 | " 8 | " " | 14 | 1.77 | 1,107 | 2.420 | 2,680 |
| 4 | " 9 | " " | 14 | 1.50 | 1,030 | 2.415 | 2,488 |
| 5 | " 10 | " " | 14 | 1.43 | 1,009 | 2.465 | 2,487 |

## APPENDIX

### DAILY GAUGE-HEIGHT [1]

U. S. WEATHER BUREAU, *Observer*

|  | 1896 | | | 1897 | | | |
|---|---|---|---|---|---|---|---|
|  | Oct. | Nov. | Dec. | Jan. | Feb. | March | April |
| 1 | Note:— Lowest water in six years was about the first of this month, — 1.20 feet. | 0.50 | 3.20 | 2.10 | 2.70 | 12.80 | 8.10 |
| 2 | | 0.10 | 6.70 | 2.10 | 3.20 | 13.50 | 10.80 |
| 3 | | 0.40 | 9.50 | 2.10 | 5.40 | 12.30 | 12.00 |
| 4 | | 1.10 | 10.70 | 2.00 | 6.20 | 9.50 | 14.00 |
| 5 | | 5.40 | 11.40 | 2.00 | 6.20 | 7.50 | 15.50 |
| 6 | | 7.70 | 12.80 | 1.90 | 6.90 | 7.00 | 15.60 |
| 7 | | 9.30 | 13.10 | 1.90 | 8.00 | 7.20 | 15.00 |
| 8 | | 10.50 | 12.40 | 1.80 | 8.70 | 8.10 | 14.80 |
| 9 | | 10.20 | 10.10 | 1.80 | 9.20 | 8.80 | 16.00 |
| 10 | | 7.50 | 7.30 | 1.80 | 9.80 | 9.60 | 16.70 |
| 11 | | 3.40 | 6.50 | 1.80 | 9.70 | 10.00 | 16.10 |
| 12 | | 2.50 | 5.60 | 1.70 | 10.80 | 10.80 | 14.80 |
| 13 | | 2.50 | 4.50 | 1.50 | 11.60 | 11.00 | 13.50 |
| 14 | | 3.60 | 4.10 | 1.50 | 13.00 | 13.00 | 12.10 |
| 15 | | 3.70 | 4.10 | 1.50 | 14.30 | 15.50 | 9.90 |
| 16 | | 3.90 | 4.00 | 1.60 | 16.10 | 20.50 | 8.00 |
| 17 | 1.10 | 3.50 | 8.00 | 4.80 | 16.00 | 22.70 | 7.20 |
| 18 | 1.10 | 2.50 | 8.90 | 5.00 | 14.60 | 21.40 | 6.80 |
| 19 | 1.10 | 1.70 | 9.70 | 4.60 | 13.10 | 20.00 | 6.40 |
| 20 | 1.20 | 1.50 | 8.10 | 5.20 | 11.70 | 18.00 | 6.60 |
| 21 | 1.20 | 1.30 | 5.30 | 6.00 | 10.20 | 16.00 | 6.40 |
| 22 | 1.20 | 1.10 | 4.10 | 6.00 | 9.20 | 14.70 | 5.00 |
| 23 | 1.20 | 1.00 | 3.70 | 7.20 | 7.60 | 15.50 | 4.50 |
| 24 | 1.20 | 0.90 | 3.20 | 7.80 | 7.50 | 16.20 | 4.40 |
| 25 | 1.10 | 0.80 | 3.00 | 8.40 | 7.80 | 17.00 | 4.30 |
| 26 | 0.60 | 0.60 | 2.80 | 8.40 | 9.90 | 17.70 | 4.10 |
| 27 | 0.20 | 0.50 | 2.50 | 6.40 | 10.50 | 17.00 | 4.00 |
| 28 | 0.10 | 0.60 | 2.30 | 5.20 | 12.00 | 15.50 | 4.00 |
| 29 | 0.20 | 0.60 | 2.30 | 3.80 | . . | 13.40 | 3.90 |
| 30 | 0.40 | 0.60 | 2.20 | 3.00 | . . | 10.80 | 4.20 |
| 31 | 0.60 | . . | 2.20 | 2.80 | . . | 8.50 | . . |

[1] In feet.

## MILLEDGEVILLE, GEORGIA

The following discharge measurements have been made at Milledgeville, on the Oconee. The section is not suitable for a regular station; but the measurements are useful, as one of them was taken at minimum stage of water. The gauge-heights are given from a bench-mark.

### DISCHARGE MEASUREMENTS

| No. | Date | Measurement Made by | Meter Number | Gauge-height in Feet | Area of Section in Square Feet | Mean Velocity in Feet per Second | Discharge in Cubic Feet per Second |
|---|---|---|---|---|---|---|---|
| 1 | 1895 Oct. 19 | C. C. Babb . . | . . . . | 1.12 | . . . . | 1.750 | 1,108 |
| 2 | 1896 Sept. 3 | Max Hall . . . | 11 | 0.70 | 344 | 1.810 | 623 |

Measurement No. 2 may be safely taken, as the minimum discharge of Oconee River at this point, as all the streams were at their lowest, at the time it was made. The important water-powers of the Oconee water-shed are reached by the Seaboard Air Line and the Georgia Railroad.

## OCMULGEE RIVER

### MACON STATION, MACON, GEORGIA

This station is at the Bibb County Highway bridge. It was established, as a station of this Survey, on October 18th, 1895, using the same rod, that the Weather Bureau had used, from 1893 to that time.

Mr. J. P. Mercer, who has been the Observer, from the time, the Survey station was established, to the present time, has been compelled, for business reasons, to resign; and Mr. W. T. Bass has been appointed in his stead.

The drainage area above Macon is 2,425 square miles:

The following is a statement of work done:—

### DISCHARGE MEASUREMENTS

| No. | Date | Measurement Made by | Meter Number | Gauge-height in Feet | Area of Section in Square Feet | Mean Velocity in Feet per Second | Discharge in Cubic Feet per Second |
|---|---|---|---|---|---|---|---|
| 1 | 1895 Oct. 18 | C. C. Babb | .. | 0.17 | ... | ... | 813 |
| 2 | Dec. 13 | " " | 62 | 1.59 | 1,045 | 1.460 | 1,530 |
| 3 | 1896 Jan. 28 | B. M. Hall | 8 | 5.52 | 2,107 | 1.630 | 3,436 |
| 4 | June 12 | " " | 8 | —0.10 | 539 | 1.470 | 791 |
| 5 | " 30 | Max Hall | 8 | —0.82 | 372 | 1.190 | 442 |
| 6 | Aug. 6 | " " | 16 | 2.97 | 1,559 | 1.230 | 2,045 |
| 7 | " 31 | " " | 11 | —0.13 | 837 | 0.776 | 651 |
| 8 | Sept. 19 | B. M. Hall | 8 | —0.85 | 625 | 0.640 | 404 |
| 9 | Oct. 16 | Max Hall | 11 | —0.61 | 667 | 0.680 | 459 |
| 10 | 1897 Mar. 15 | B. M. Hall | 91 | 16.75 | 5,862 | 4.356 | 25,535 |
| 11 | May 4 | " " | 91 | 4.30 | 1,612 | 1.706 | 2,750 |
| 12 | " 5 | " " | 91 | 3.50 | 1,412 | 1.623 | 2,275 |
| 13 | " 18 | Max Hall | 11 | 2.10 | 1,092 | 1.458 | 1,592 |
| 14 | June 11 | P. A. Dallis | 14 | 2.85 | 1,325 | 1.590 | 2,111 |
| 15 | " 12 | " " | 14 | 1.85 | 1,045 | 1.415 | 1,479 |
| 16 | " 29 | B. M. Hall | 91 | 0.90 | 829 | 1.213 | 1,005 |

## Macon Station — *Continued*

### DAILY GAUGE-HEIGHT[1]

J. P. Mercer, *Observer*

| 1895 | | | | | | | |
|---|---|---|---|---|---|---|---|
| | October | November | December | | October | November | December |
| 1  | . . | 0.50 | 0.50 | 17 | . .  | 0.57 | 0.58 |
| 2  | . . | 0.77 | 0.50 | 18 | . .  | 0.55 | 0.64 |
| 3  | . . | 0.85 | 0.55 | 19 | . .  | 0.50 | 0.61 |
| 4  | . . | 0.67 | 0.62 | 20 | . .  | 0.50 | 0.59 |
| 5  | . . | 0.55 | 0.54 | 21 | . .  | 0.50 | 2.02 |
| 6  | . . | 0.45 | 0.51 | 22 | . .  | 0.50 | 3.10 |
| 7  | . . | 0.36 | 0.44 | 23 | 0.21 | 0.50 | 2.68 |
| 8  | . . | 0.47 | 0.40 | 24 | 0.21 | 0.50 | 1.70 |
| 9  | . . | 0.55 | 0.46 | 25 | 0.17 | 0.49 | 1.48 |
| 10 | . . | 0.63 | 0.45 | 26 | 0.19 | 0.49 | 1.01 |
| 11 | . . | 0.65 | 2.50 | 27 | 0.18 | 0.47 | 1.00 |
| 12 | . . | 0.60 | 2.29 | 28 | 0.18 | 0.43 | 1.20 |
| 13 | . . | 0.77 | 1.51 | 29 | 0.17 | 0.55 | 1.30 |
| 14 | . . | 0.94 | 1.11 | 30 | 0.22 | 0.54 | 1.35 |
| 15 | . . | 0.72 | 1.01 | 31 | 0.50 | . .  | 4.46 |
| 16 | . . | 0.65 | 0.72 |    |      |      |      |

[1] In feet.

## APPENDIX

### DAILY GAUGE-HEIGHT—*Continued* [1]

J. P. MERCER, *Observer*

| | \multicolumn{12}{c}{1896} |
|---|---|---|---|---|---|---|---|---|---|---|---|---|
| | Jan. | Feb. | Mar. | April | May | June | July | Aug. | Sept. | Oct. | Nov. | Dec. |
| 1 | 4.81 | 3.00 | 2.50 | 3.10 | 0.89 | 0.08 | —0.90 | 1.02 | 0.11 | —0.82 | —0.14 | 9.50 |
| 2 | 4.50 | 2.09 | 2.20 | 3.00 | 0.75 | 0.11 | —1.00 | 1.50 | 0.12 | —0.86 | —0.08 | 11.08 |
| 3 | 2.20 | 3.20 | 2.10 | 4.60 | 0.63 | 0.56 | 0.10 | 1.82 | 0.11 | —0.88 | —0.02 | 12.60 |
| 4 | 1.70 | 3.00 | 2.00 | 4.40 | 0.90 | 0.85 | 0.15 | 2.22 | 0.28 | —0.81 | 10.00 | 10.20 |
| 5 | 1.42 | 2.09 | 2.00 | 2.90 | 2.12 | 1.92 | 2.00 | 2.62 | 0.19 | —0.75 | 14.20 | 8.00 |
| 6 | 1.08 | 13.50 | 1.90 | 1.80 | 2.73 | 1.52 | 4.00 | 3.00 | 0.19 | —0.79 | 14.40 | 6.15 |
| 7 | 0.96 | 10.70 | 4.00 | 1.60 | 1.97 | 0.96 | 5.30 | 3.05 | 0.19 | —0.78 | 8.80 | 4 52 |
| 8 | 1.72 | 7.50 | 6.00 | 1.10 | 1.62 | 0.73 | 11.00 | 2.78 | 0.12 | —0.82 | 5.40 | 3.62 |
| 9 | 2.83 | 13.10 | 5.00 | 1.10 | 0.86 | 0.25 | 19.70 | 2.41 | 0.08 | —0.82 | 3.22 | 3.00 |
| 10 | 2.77 | 11.30 | 4.00 | 1.40 | 0.61 | 0.01 | 19.40 | 1.88 | 0.04 | —0.82 | 2.25 | 2.42 |
| 11 | 2.10 | 8.70 | 5.00 | 1.30 | 0.43 | —0.05 | 15.00 | 1.48 | —0.01 | —0.73 | 1.50 | 1.98 |
| 12 | 1.60 | 7.00 | 7.20 | 1.10 | 0.30 | —0.10 | 10.20 | 0.40 | —0.01 | —0.65 | 1.18 | 1.58 |
| 13 | 1.50 | 6.30 | 6.50 | 1.20 | 0.19 | —0.17 | 8.20 | 0.20 | —0.31 | —0.75 | 10.00 | 1.26 |
| 14 | 1.20 | 6.50 | 6.20 | 1.20 | 0.11 | 0.29 | 7.10 | 0.13 | —0.45 | —0.73 | 8.10 | 1.12 |
| 15 | 2.00 | 5.00 | 6.00 | 1.13 | 0.09 | 0.32 | 7.00 | 0.25 | —0.80 | —0.65 | 5.32 | 11.70 |
| 16 | 2.50 | 4.80 | 6.00 | 1.12 | 0.07 | 0.20 | 6.20 | 1.08 | —0.68 | —0.77 | 1.53 | 6.00 |
| 17 | 7.20 | 4.20 | 5.50 | 1.09 | 0.05 | 0.25 | 16.00 | 0.50 | —0.78 | —0.80 | 1.14 | 4.62 |
| 18 | 5.00 | 3.40 | 5.30 | 1.05 | 0.03 | 0.25 | 18.20 | 0.60 | —0.80 | —0.83 | 0.97 | 3.94 |
| 19 | 4.50 | 3.20 | 5.00 | 0.98 | —0.05 | 0.30 | 13.00 | 0.38 | —0.82 | —0.85 | 0.85 | 2.85 |
| 20 | 4.00 | 3.00 | 4.90 | 0.94 | —0.10 | 0.40 | 7.05 | 0.20 | —0.91 | —0.88 | 0.63 | 2.38 |
| 21 | 3.90 | 2.90 | 4.70 | 0.86 | —0.10 | 0.47 | 3.80 | 0.11 | —0.90 | —0.89 | 0.70 | 2.00 |
| 22 | 3.70 | 2.70 | 4.50 | 0.76 | —0.15 | 0.56 | 3.20 | 0.08 | —0.82 | —0.90 | 0.62 | 1.90 |
| 23 | 9.40 | 2.60 | 4.20 | 0.71 | —0.05 | 0.70 | 3.00 | 0.06 | —0.41 | —0.77 | 0.58 | 1.76 |
| 24 | 13.80 | 2.50 | 5.00 | 0.63 | 0.56 | 0.50 | 2.90 | 0.05 | —0.48 | —0.40 | 0.58 | 1.38 |
| 25 | 12.00 | 2.48 | 5.00 | 0.63 | 0.50 | 0.70 | 2.85 | 0.00 | —0.61 | —0.52 | 0.51 | 1.18 |
| 26 | 9.30 | 2.40 | 4.80 | 1.02 | 0.78 | —0.20 | 2.60 | 0.03 | —0.72 | —0.25 | 0.51 | 1.11 |
| 27 | 7.00 | 2.30 | 4.70 | 2.90 | 0.52 | —0.35 | 2.40 | 0.04 | —0.78 | —0.08 | 0.50 | 0.96 |
| 28 | 5.80 | 3.30 | 4.60 | 2.32 | 0.34 | —0.35 | 2.10 | 0.07 | —0.83 | —0.20 | 0.47 | 0.90 |
| 29 | 5.30 | 2.80 | 4.40 | 1.36 | 0.17 | —0.75 | 1.92 | 0.07 | —0.71 | —0.32 | 0.44 | 0.83 |
| 30 | 4.80 | . . | 4.20 | 1.02 | 0.12 | —0.85 | 1.60 | 0.09 | —0.80 | —0.23 | 0.32 | 0.78 |
| 31 | 3.20 | . . | 4.10 | . . | 0.12 | . . . | 1.41 | 0.12 | . . . | —0.19 | . . . | 0.70 |

[1] In feet.

## DAILY GAUGE-HEIGHT — *Continued* [1]

### J. P. MERCER, *Observer*

| | \multicolumn{6}{c}{1897} | | | | | | |
|---|---|---|---|---|---|---|---|---|---|---|---|---|
| | Jan. | Feb. | Mar. | Apr. | May | June | | Jan. | Feb. | Mar. | Apr. | May | June |
| 1 | 0.68 | 2.00 | 4.00 | 5.57 | 1.95 | 1.22 | 17 | 0.44 | 5.12 | 9.45 | 3.70 | 2.50 | 1.52 |
| 2 | 0.63 | 5.00 | 3.70 | 9.75 | 1.90 | 1.18 | 18 | 1.15 | 2.00 | 8.25 | 3.50 | 2.22 | 1.45 |
| 3 | 0.60 | 8.00 | 3.20 | 10.05 | 1.87 | 3.15 | 19 | 1.50 | 2.75 | 8.20 | 3.20 | 2.09 | 1.37 |
| 4 | 0.58 | 6.00 | 2.00 | 10.00 | 2.15 | 3.20 | 20 | 2.10 | 2.62 | 9.57 | 3.00 | 1.84 | 3.25 |
| 5 | 0.56 | 6.00 | 4.00 | 15.12 | 2.23 | 3.12 | 21 | 1.25 | 2.65 | 10.00 | 2.90 | 1.81 | 3.12 |
| 6 | 0.54 | 8.00 | 3.00 | 15.15 | 2.47 | 3.10 | 22 | 7.00 | 2.71 | 9.00 | 2.80 | 1.78 | 2.80 |
| 7 | 0.52 | 6.50 | 11.60 | 12.60 | 3.00 | 3.00 | 23 | 5.50 | 2.00 | 15.50 | 2.70 | 1.71 | 2.62 |
| 8 | 0.50 | 5.00 | 12.70 | 10.48 | 3.00 | 2.54 | 24 | 3.00 | 1.96 | 14.00 | 2.60 | 1.68 | 2.70 |
| 9 | 0.49 | 4.75 | 7.50 | 10.00 | 2.91 | 2.32 | 25 | 2.25 | 6.00 | 10.60 | 2.40 | 1.60 | 3.00 |
| 10 | 0.49 | 4.55 | 5.00 | 10.80 | 2.72 | 2.26 | 26 | 2.00 | 10.50 | 8.30 | 2.30 | 1.56 | 3.11 |
| 11 | 0.51 | 5.00 | 4.80 | 7.80 | 2.57 | 2.18 | 27 | 3.00 | 7.02 | 7.10 | 2.22 | 1.53 | 2.50 |
| 12 | 0.49 | 13.50 | 6.00 | 6.40 | 2.45 | 2.08 | 28 | 3.12 | 5.00 | 6.20 | 2.16 | 1.50 | 1.00 |
| 13 | 0.48 | 12.75 | 17.30 | 5.00 | 3.05 | 2.04 | 29 | 1.50 | . . | 5.57 | 2.08 | 1.47 | 0.90 |
| 14 | 0.53 | 7.00 | 18.00 | 4.70 | 3.15 | 2.01 | 30 | 1.25 | . . | 5.21 | 2.00 | 1.36 | 1.50 |
| 15 | 0.46 | 5.00 | 17.70 | 4.00 | 3.28 | 1.89 | 31 | 1.20 | . . | 5.20 | . . | 1.28 | . . |
| 16 | 0.44 | 5.00 | 13.00 | 4.00 | 3.00 | 1.73 | | | | | | | |

[1] In feet.

## APPENDIX

### Macon Station — *Continued*

#### RATING-TABLE

*Drainage Area, 2,425 Square Miles*

| Gauge-height in Feet | Discharge in Cubic Feet per Second | Gauge-height in Feet | Discharge in Cubic Feet per Second | Gauge-height in Feet | Discharge in Cubic Feet per Second | Gauge-height in Feet | Discharge in Cubic Feet per Second | Gauge-height in Feet | Discharge in Cubic Feet per Second |
|---|---|---|---|---|---|---|---|---|---|
| .. | ... | 1.00 | 1,200 | 3.00 | 2,050 | 5.00 | 3,090 | 7.00 | 4,600 |
| —0.85 | 404 | 1.10 | 1,242 | 3.10 | 2,100 | 5.10 | 3,130 | 8.00 | 5,750 |
| —0.80 | 426 | 1.20 | 1,285 | 3.20 | 2,150 | 5.20 | 3,210 | 9.00 | 7,250 |
| —0.70 | 469 | 1.30 | 1,328 | 3.30 | 2,195 | 5.30 | 3,275 | 10.00 | 8,625 |
| —0.60 | 512 | 1.40 | 1,371 | 3.40 | 2,240 | 5.40 | 3,340 | 11.00 | 10,300 |
| —0.50 | 555 | 1.50 | 1,414 | 3.50 | 2,285 | 5.50 | 3,400 | 12.00 | 11,975 |
| —0.40 | 598 | 1.60 | 1,457 | 3.60 | 2,330 | 5.60 | 3,460 | 13.00 | 14,000 |
| —0.30 | 641 | 1.70 | 1,500 | 3.70 | 2,375 | 5.70 | 3,530 | 14.00 | 16,750 |
| —0.20 | 684 | 1.80 | 1,543 | 3.80 | 2,420 | 5.80 | 3,600 | 15.00 | 19,750 |
| —0.10 | 727 | 1.90 | 1,586 | 3.90 | 2,470 | 5.90 | 3,675 | 16.00 | 23,000 |
| 0.00 | 770 | 2.00 | 1,629 | 4.00 | 2,520 | 6.00 | 3,750 | 16.75 | 25,535 |
| 0.10 | 813 | 2.10 | 1,672 | 4.10 | 2,575 | 6.10 | 3,825 | 17.00 | 26,200 |
| 0.20 | 855 | 2.20 | 1,715 | 4.20 | 2,630 | 6.20 | 3,900 | 18.00 | 29,375 |
| 0.30 | 898 | 2.30 | 1,758 | 4.30 | 2,685 | 6.30 | 3,985 | 19.00 | 32,750 |
| 0.40 | 941 | 2.40 | 1,801 | 4.40 | 2,740 | 6.40 | 4,070 | 19.70 | 35,150 |
| 0.50 | 984 | 2.50 | 1,844 | 4.50 | 2,800 | 6.50 | 4,155 | 20.00 | 36,200 |
| 0.60 | 1,027 | 2.60 | 1,887 | 4.60 | 2,860 | 6.60 | 4,240 | .. | ... |
| 0.70 | 1,070 | 2.70 | 1,920 | 4.70 | 2,915 | 6.70 | 4,335 | .. | ... |
| 0.80 | 1,113 | 2.80 | 1,963 | 4.80 | 2,970 | 6.80 | 4,430 | .. | ... |
| 0.90 | 1,156 | 2.90 | 2,006 | 4.90 | 3,030 | 6.90 | 4,515 | .. | ... |

# YELLOW RIVER

### Almon, Newton County, Georgia

Macon is the only regular station, on the Ocmulgee water-shed; but the following discharge measurements have been made on YELLOW RIVER, AT ALMON, NEWTON COUNTY, at the wagon bridge, just below the Georgia Railroad bridge. A rod has been set there, for the comparison of different measurements.

#### DISCHARGE MEASUREMENTS

| No. | Date | Measurement Made by | Meter Number | Gauge-height in Feet | Area of Section in Square Feet | Mean Velocity in Feet per Second | Discharge in Cubic Feet per Second |
|---|---|---|---|---|---|---|---|
| 1 | 1896 Sept. 19 | Max Hall | 11 | 0.75 | 38 | 1.63 | 62.4 |
| 2 | 1897 Mar. 27 | B. M. Hall | 91 | 3.91 | 469 | 1.86 | 876.0 |
| 3 | 1897 June 21 | Max Hall | 91 | 2.50 | 305 | 0.94 | 287.0 |

Measurement No. 1 was made, at the time of lowest water; but, as there are mill-ponds above, it, probably, does not represent the full volume of the river. The numerous water-powers of the Ocmulgee water-shed are reached by the Southern Railway System and the Georgia Railroad.

# APALACHICOLA BASIN

## FLINT RIVER

### Molena Station, Molena, Georgia

This station, which is at the bridge of the Georgia Midland Division of the Southern Railway, on the line of Pike and Meriwether counties, was established May 21st, 1897.

The gauge-observer, Mr. J. A. Moore, began his duties June 7th; but the list of gauge-heights is, thus far, too short to publish.

DISCHARGE MEASUREMENTS

| No | Date | Measurement Made by | Meter Number | Gauge-height in Feet | Area of Section in Square Feet | Mean Velocity in Feet per Second | Discharge in Cubic Feet per Second |
|---|---|---|---|---|---|---|---|
| 1 | 1897 May 21 | B. M. Hall | 91 | 1.50 | 791 | 0.810 | 641 |
| 2 | June 7 | Max Hall | 91 | 1.75 | 869 | 0.815 | 707 |
| 3 | " 23 | B. M. Hall | 91 | 1.70 | 837 | 0.832 | 697 |

On June 23rd, 1897, a discharge measurement was, also, made on Red Oak creek, at its mouth, which creek is a large tributary, entering Flint river, about three miles above Molena Station. Its discharge was found to be 101 cubic feet per second, while that of Flint river, at Molena Station, was 697 cubic feet per second.

## Reynolds, Georgia

The only other discharge measurement, made on the Flint river, was at Reynolds, where the Central Railroad crosses; and it is as follows: —

### DISCHARGE MEASUREMENT

| No. | Date | Measurement Made by | Meter Number | Gauge-height in Feet | Area of Section in Square Feet | Mean Velocity in Feet per Second | Discharge in Cubic Feet per Second |
|---|---|---|---|---|---|---|---|
| 1 | 1897 June 11 | B. M. Hall | 14 | 0.95 [1] | 1,332 | 1.36 | 1,810 |

[1] On Weather Bureau rod.

## CHATTAHOOCHEE RIVER

### OAKDALE STATION, FULTON COUNTY, GEORGIA

Oakdale Station, at the bridge of the Georgia Pacific Division of the Southern Railway, in Fulton county, was established October 15th, 1895, with Mr. J. H. Lowry, as gauge-observer. The drainage area above Oakdale Station is 1,560 square miles. The following is a statement of work done at this station:—

DISCHARGE MEASUREMENTS

| No. | Date | Measurement Made by | Meter Number | Gauge-height in Feet | Area of Section in Square Feet | Mean Velocity in Feet per Second | Discharge in Cubic Feet per Second |
|---|---|---|---|---|---|---|---|
| | 1895 | | | | | | |
| 1 | Oct. 15 | C. C. Babb | ... | 0.40 | ..... | ..... | 1,096 |
| 2 | Dec. 14 | " " | 62 | 0.69 | ..... | ..... | 1,380 |
| | 1896. | | | | | | |
| 3 | Jan. 14 | B. M. Hall | 8 | 0.70 | 888 | 1.530 | 1,361 |
| 4 | June 15 | " " | 8 | 0.00 | 704 | 1.400 | 985 |
| 5 | " 20 | " " | 8 | 0.33 | 792 | 1.450 | 1,153 |
| 6 | " 22 | " " | 8 | 1.01 | ..... | ..... | 1,530 |
| 7 | " 23 | " " | 8 | 0.55 | 841 | 1.480 | 1,250 |
| 8 | " 24 | Max Hall | 8 | 0.28 | 729 | 1.540 | 1,126 |
| 9 | July 9 | " " | 8 | 18.05 | ..... | ..... | 24,100 |
| 10 | " 10 | B. M. Hall | 8 | 12.80 | ..... | ..... | 16,200 |
| 11 | " 13 | Max Hall | 8 | 3.01 | 1,161 | 2.550 | 2,957 |
| 12 | " 15 | " " | 8 | 1.88 | 961 | 2.150 | 2,066 |
| 13 | " 17 | " " | 8 | 4.61 | 1,471 | 3.150 | 4,640 |
| 14 | " 24 | " " | 8 | 2.22 | 1,028 | 2.400 | 2,470 |
| 15 | Aug. 29 | " " | 11 | —0.18 | 517 | 1.880 | 958 |
| 16 | Sept. 9 | " " | 11 | —1.55 | 422 | 1.760 | 744 |
| 17 | Oct. 17 | " " | 8 | —0.50 | 420 | 1.840 | 775 |
| | 1897 | | | | | | |
| 18 | Apr. 24 | " " | 91 | 2.90 | 1,244 | 2.520 | 3,065 |
| 19 | " 27 | " " | 16 | 2.70 | 1,164 | 2.320 | 2,703 |

## DISCHARGE MEASUREMENTS — *Continued*

| No. | Date | Measurement Made by | Meter Number | Gauge-height in Feet | Area of Section in Square Feet | Mean Velocity in Feet per Second | Discharge in Cubic Feet per Second |
|---|---|---|---|---|---|---|---|
| 20 | 1897 May 22. | B. M. Hall | 91 | 1.65 | 873 | 2.350 | 2,055 |
| 21 | " 25. | " " | 14 | 1.50 | 911 | 2.200 | 2,014 |
| 22 | " 31. | Max Hall | 91 | 1.35 | 844 | 2.373 | 2,003 |
| 23 | " 31. | " " | 14 | 1.35 | 844 | 2.283 | 1,927 |
| 24 | June 9. | B. M. Hall | 91 | 1.44 | 889 | 2.240 | 1,991 |
| 25 | " 16. | P. A. Dallis | 14 | 0.94 | 831 | 1.833 | 1,523 |
| 26 | " 28. | Max Hall | 70 | 0.57 | 676 | 1.931 | 1,306 |

## DAILY GAUGE-HEIGHT [1]

### J. H. LOWRY, *Observer*

| 1895 | | | | | | | |
|---|---|---|---|---|---|---|---|
| | October | November | December | | October | November | December |
| 1 | . . | 0.75 | 0.50 | 17 | 0.35 | 0.55 | 0.55 |
| 2 | . . | 1.70 | 0.50 | 18 | 0.30 | 0.60 | 0.50 |
| 3 | . . | 1.00 | 0.55 | 19 | 0.25 | 0.50 | 0.40 |
| 4 | . . | 0.60 | 0.60 | 20 | 0.30 | 0.55 | 0.50 |
| 5 | . . | 0.50 | 0.60 | 21 | 0.20 | 0.50 | 0.65 |
| 6 | . . | 0.55 | 0.50 | 22 | 0.25 | 0.50 | 1.00 |
| 7 | . . | 0.45 | 0.45 | 23 | 0.20 | 0.50 | 2.00 |
| 8 | . . | 0.40 | 0.40 | 24 | 0.30 | 0.50 | 1.20 |
| 9 | . . | 0.60 | 0.40 | 25 | 0.25 | 0.50 | 1.00 |
| 10 | . . | 0.80 | 0.60 | 26 | 0.20 | 0.55 | 0.55 |
| 11 | . . | 1.00 | 1.30 | 27 | 0.25 | 0.55 | 0.80 |
| 12 | . . | 1.35 | 1.10 | 28 | 0.25 | 0.70 | 0.75 |
| 13 | . . | 1.00 | 1.00 | 29 | 0.30 | 0.75 | 0.70 |
| 14 | . . | 0.75 | 0.65 | 30 | 0.50 | 0.45 | 2.00 |
| 15 | 0.40 | 0.60 | 0.60 | 31 | 0.50 | . . | 2.95 |
| 16 | 0.40 | 0.60 | 0.55 | | | | |

[1] In feet.

## APPENDIX

### DAILY GAUGE-HEIGHT — *Continued* [1]

J. H. LOWRY, *Observer*

| 1896 | Jan. | Feb. | Mar. | April | May | June | July | Aug. | Sept. | Oct. | Nov. | Dec. |
|---|---|---|---|---|---|---|---|---|---|---|---|---|
| 1  | 3.20 | 1.70 | 1.40 | 1.40 | 0.75 | 0.25  | —0.35 | 0.45 | —0.40 | 0.20  | 0.50 | 0.50 |
| 2  | 2.00 | 1.50 | 1.35 | 1.70 | 0.75 | 1.40  | —0.40 | 0.50 | —0.60 | 0.10  | 0.50 | 1.10 |
| 3  | 1.00 | 1.75 | 1.40 | 2.75 | 0.75 | 1.45  | —0.45 | 1.40 | —0.60 | 0.05  | 0.65 | 1.30 |
| 4  | 1.20 | 2.10 | 1.45 | 1.85 | 0.70 | 2.10  | 0.15  | 1.70 | —0.60 | —0.20 | 1.00 | 3.20 |
| 5  | 1.10 | 2.20 | 1.40 | 1.35 | 0.65 | 1.60  | 0.20  | 0.75 | —0.60 | —0.30 | 1.60 | 2.30 |
| 6  | 1.00 | 2.80 | 1.35 | 1.30 | 3.15 | 1.15  | —0.15 | 0.40 | —0.60 | —0.35 | 1.40 | 1.30 |
| 7  | 0.80 | 3.70 | 1.95 | 1.10 | 1.70 | 0.60  | 2.00  | 0.20 | —0.60 | —0.40 | 1.20 | 1.15 |
| 8  | 0.95 | 4.30 | 1.75 | 1.15 | 1.20 | 0.40  | 12.25 | 0.15 | —0.55 | —0.50 | 1.10 | 1.00 |
| 9  | 1.20 | 5.80 | 1.55 | 1.05 | 0.90 | 0.70  | 17.70 | 0.15 | —0.55 | —0.50 | 1.00 | 1.00 |
| 10 | 1.25 | 6.55 | 1.50 | 1.05 | 0.75 | 0.70  | 18.45 | 0.10 | —0.50 | —0.50 | 1.05 | 1.00 |
| 11 | 1.00 | 4.30 | 1.75 | 1.05 | 0.65 | 0.65  | 4.75  | 0.05 | —0.50 | —0.60 | 1.05 | 0.90 |
| 12 | 0.85 | 3.30 | 1.85 | 1.00 | 0.60 | 0.50  | 3.25  | 0.05 | —0.55 | —0.60 | 1.20 | 0.70 |
| 13 | 0.80 | 2.80 | 1.55 | 1.00 | 0.50 | 0.30  | 3.10  | 0.30 | —0.60 | —0.55 | 3.80 | .60 |
| 14 | 0.75 | 2.95 | 1.35 | 0.90 | 0.50 | 0.20  | 3.90  | 0.25 | —0.60 | —0.55 | 4.60 | .65 |
| 15 | 0.70 | 4.10 | 1.20 | 0.90 | 0.40 | —0.05 | 1.95  | 0.05 | —0.60 | —0.55 | 2.60 | .70 |
| 16 | 0.85 | 2.90 | 1.25 | 0.90 | 0.30 | 0.05  | 3.00  | 0.10 | —0.65 | —0.55 | 1.50 | .65 |
| 17 | 2.40 | 2.55 | 1.10 | 0.85 | 0.30 | —0.05 | 4.40  | 0.10 | —0.65 | —0.55 | 1.00 | .65 |
| 18 | 2.35 | 2.20 | 1.35 | 0.85 | 0.25 | 0.10  | 2.70  | 0.10 | —6.65 | —0.55 | 0.80 | .65 |
| 19 | 2.10 | 2.10 | 1.70 | 0.85 | 0.25 | 0.15  | 1.75  | —0.10 | —0.65 | —0.55 | 0.60 | .65 |
| 20 | 1.50 | 1.80 | 1.65 | 0.85 | 0.20 | 0.55  | 1.90  | —0.25 | —0.65 | —0.55 | 0.50 | .65 |
| 21 | 1.25 | 1.60 | 1.45 | 0.75 | 0.15 | 1.60  | 1.65  | —0.30 | —0.50 | —0.55 | 0.40 | .60 |
| 22 | 1.80 | 1.45 | 1.35 | 0.75 | 0.20 | 0.45  | 1.90  | —0.30 | —0.15 | —0.50 | 0.40 | .55 |
| 23 | 6.30 | 1.55 | 1.40 | 0.65 | 0.20 | 0.50  | 2.45  | —0.30 | 0.40  | —0.35 | 0.30 | .50 |
| 24 | 9.80 | 1.60 | 1.45 | 0.70 | 1.10 | 0.30  | 2.40  | —0.30 | 0.60  | —0.05 | 0.25 | .50 |
| 25 | 9.95 | 1.50 | 1.40 | 0.65 | 1.55 | 0.20  | 1.50  | 1.00  | —0.10 | 0.05  | 0.20 | .45 |
| 26 | 5.10 | 1.40 | 1.35 | 0.65 | 0.65 | —0.10 | 1.30  | —0.10 | —0.40 | 0.10  | 0.20 | .45 |
| 27 | 3.55 | 1.35 | 1.25 | 1.40 | 0.85 | 0.45  | 0.90  | —0.30 | —0.40 | 0.15  | 0.20 | .40 |
| 28 | 2.65 | 1.30 | 1.20 | 1.70 | 0.65 | —0.10 | 0.80  | —0.30 | 0.20  | 0.25  | 0.20 | .35 |
| 29 | 2.30 | 1.55 | 1.25 | 1.15 | 0.95 | —0.20 | 0.80  | —0.30 | 0.45  | 0.90  | 0.20 | .35 |
| 30 | 2.10 | ... | 1.20 | 0.85 | 0.40 | —0.25 | 0.50  | —0.35 | 0.30  | 0.65  | 0.20 | .30 |
| 31 | 1.80 | ... | 1.20 | ... | 0.35 | ... | —0.40 | ... | | 0.50 | . | .25 |

[1] In Feet.

## DAILY GAUGE-HEIGHT — *Continued* [1]

### J. H. LOWRY, *Observer*

| | \multicolumn{6}{c}{1897} | | | | | | |
|---|---|---|---|---|---|---|---|---|---|---|---|---|
| | Jan. | Feb. | Mar. | Apr. | May | June | | Jan. | Feb. | Mar. | Apr. | May | June |
| 1 | 0.20 | 3.10 | 2.00 | 2.80 | 4.10 | 1.00 | 17 | 2.40 | 2.40 | 5.45 | 3.50 | 1.70 | 0.55 |
| 2 | 0.20 | 4.50 | 1.80 | 4.10 | 3.20 | 1.00 | 18 | 4.00 | 2.10 | 6.10 | 2.20 | 1.50 | 1.00 |
| 3 | 0.20 | 4.90 | 1.80 | 5.00 | 2.80 | 1.15 | 19 | 4.60 | 1.90 | 5.80 | 3.05 | 1.45 | 1.05 |
| 4 | 0.30 | 3.25 | 2.00 | 7.00 | 2.50 | 1.35 | 20 | 5.50 | 2.50 | 5.55 | 3.00 | 1.35 | 0.55 |
| 5 | 0.40 | 3.70 | 1.90 | 12.80 | 2.40 | 1.15 | 21 | 7.50 | 2.50 | 5.00 | 2.80 | 1.35 | 1.35 |
| 6 | 0.40 | 4.10 | 9.00 | 17.00 | 2.25 | 1.05 | 22 | 6.85 | 2.90 | 4.30 | 2.75 | 1.40 | 0.75 |
| 7 | 0.35 | 5.50 | 9.20 | 8.00 | 2.15 | 1.00 | 23 | 3.90 | 5.10 | 4.50 | 2.70 | 1.40 | 0.55 |
| 8 | 0.35 | 5.00 | 5.50 | 5.40 | 2.05 | 0.85 | 24 | 3.20 | 4.50 | 4.00 | 2.70 | 1.35 | 0.45 |
| 9 | 0.30 | 3.10 | 4.45 | 6.00 | 2.00 | 1.10 | 25 | 2.00 | 3.60 | 3.90 | 2.65 | 1.30 | 0.90 |
| 10 | 0.30 | 2.40 | 4.05 | 5.00 | 1.95 | 1.00 | 26 | 1.65 | 2.90 | 3.35 | 2.60 | 1.20 | 0.55 |
| 11 | 0.20 | 3.40 | 3.60 | 4.60 | 1.90 | 1.00 | 27 | 0.45 | 2.40 | 3.10 | 2.40 | 1.10 | 0.45 |
| 12 | 2.20 | 4.00 | 6.40 | 4.10 | 1.90 | 0.85 | 28 | 1.10 | 2.05 | 3.00 | 2.10 | 1.05 | 0.30 |
| 13 | 0.40 | 5.30 | 12.60 | 4.00 | 1.85 | 0.75 | 29 | 1 25 | . . | 2.75 | 2.00 | 1.00 | 1.00 |
| 14 | 3.00 | 4.00 | 10.00 | 4.55 | 1.90 | 0.65 | 30 | 1.50 | . . | 2.70 | 2.20 | 1.05 | 0.75 |
| 15 | 3.70 | 2.60 | 8.40 | 4.05 | 2.35 | 0.55 | 31 | 2.00 | . . | 2.15 | . . | 1.10 | . . |
| 16 | 2.80 | 2.60 | 6.80 | 4.00 | 1.85 | 0.50 | | | | | | | |

[1] In inches.

## RATING-TABLE

*Drainage Area, 1560 Square Miles*

| Gauge-height in feet | Discharge in Cubic Feet per Second | Gauge-height in Feet | Discharge in Cubic Feet per Second | Gauge-height in Feet | Discharge in Cubic Feet per Second | Gauge-height in Feet | Discharge in Cubic Feet per Second |
|---|---|---|---|---|---|---|---|
| . . | . . | 1.00 | 1,528 | 3.00 | 2,956 | 5.00 | 5,170 |
| . . | . . | 1.10 | 1,586 | 3.10 | 3,044 | . . | . . |
| . . | . . | 1.20 | 1,646 | 3.20 | 3,133 | . . | . . |
| . . | . . | 1.30 | 1,707 | 3.30 | 3,223 | . . | . . |
| —0.55 | 744 | 1.40 | 1,769 | 3.40 | 3,315 | . . | . . |
| —0.50 | 775 | 1.50 | 1,832 | 3.50 | 3,410 | . . | . . |
| —0.40 | 821 | 1.60 | 1,896 | 3.60 | 3,508 | . . | . . |
| —0.30 | 856 | 1.70 | 1,961 | 3.70 | 3,608 | . . | . . |
| —0.20 | 895 | 1.80 | 2,027 | 3.80 | 3,711 | . . | . . |
| —0.10 | 938 | 1.90 | 2,085 | 3.90 | 3,817 | . . | . . |
| 0.00 | 985 | 2.00 | 2,155 | 4.00 | 3,928 | . . | . . |
| 0.10 | 1,035 | 2.10 | 2,227 | 4.10 | 4,040 | . . | . . |
| 0.20 | 1,086 | 2.20 | 2,301 | 4.20 | 4,154 | . . | . . |
| 0.30 | 1,138 | 2.30 | 2,377 | 4.30 | 4,271 | . . | . . |
| 0.40 | 1,191 | 2.40 | 2,455 | 4.40 | 4,391 | . . | . . |
| 0.50 | 1,245 | 2.50 | 2,535 | 4.50 | 4,514 | . . | . . |
| 0.60 | 1,300 | 2.60 | 2,616 | 4.60 | 4,640 | . . | . . |
| 0.70 | 1,356 | 2.70 | 2,698 | 4.70 | 4,768 | . . | . . |
| 0.80 | 1,412 | 2.80 | 2,782 | 4.80 | 4,899 | . . | . . |
| 0.90 | 1,469 | 2.90 | 2,868 | 4.90 | 5,033 | . . | . . |

The minimum discharge, per square mile of drainage area, is 0.48 cubic feet per second.

## West Point Station, West Point, Georgia

The station at West Point was established July 30th, 1896, at the iron highway bridge, though one measurement was made by Mr. C. C. Babb, of the U. S. Geological Survey, in October, 1895. Mr. C. E. Melton was appointed gauge-observer. The drainage area above this point is 3,300 square miles. The following statement shows the work done at this station: —

### DISCHARGE MEASUREMENTS

| No. | Date | Measurement Made by | Meter Number | Gauge-height in Feet | Area of Section in Square Feet | Mean Velocity in Feet per Second | Discharge in Cubic Feet per Second |
|---|---|---|---|---|---|---|---|
| 1 | 1895 Oct. 22 | C. C. Babb | 76 | 1.76 | 2,802 | 0.510 | 1,404 |
| 2 | 1896 July 30 | Max Hall | 16 | 2.45 | 3,249 | 0.748 | 2,430 |
| 3 | Aug. 14 | " " | 16 | 1.72 | 3,077 | 0.515 | 1,594 |
| 4 | Sept. 5 | " " | 11 | 1.20 | . . | 0.352 | 1,050 |
| 5 | " 25 | B. M. Hall | 8 | 1.15 | 2,792 | 0.370 | 1,030 |
| 6 | Oct. 28 | Max Hall | 11 | 1.75 | 2,883 | 0.570 | 1,642 |
| 7 | 1897 Jan. 23 | B. M. Hall | 11 | 6.66 | 4,597 | 2.593 | 1,192 |
| 8 | Apr. 26 | Max Hall | 91 | 3.70 | 3,855 | 1.413 | 5,448 |
| 9 | May 4 | " " | 11 | 4.13 | 4,082 | 1.526 | 6,230 |
| 10 | " 19 | " " | 91 | 3.00 | 3,556 | 1.000 | 3,557 |
| 11 | June 5 | " " | 14 | 2.90 | 3,552 | 0.915 | 3,253 |
| 12 | " 29 | " " | 91 | 2.59 | 3,407 | 0.861 | 2,934 |

## DAILY GAUGE-HEIGHT [1]

C. E. MELTON, *Observer*

| | \multicolumn{5}{c}{1896} | | | | | | |
|---|---|---|---|---|---|---|---|---|---|---|---|
| | Aug. | Sept. | Oct. | Nov. | Dec. | | Aug. | Sept. | Oct. | Nov. | Dec. |
| 1 | 2.70 | 1.30 | 4.10 | 1.70 | 4.20 | 17 | 1.60 | 0.85 | 1.10 | 3.00 | 3.00 |
| 2 | 3.90 | 1.20 | 4.00 | 2.00 | 4.00 | 18 | 1.55 | 0.80 | 1.10 | 2.60 | 3.00 |
| 3 | 4.50 | 1.10 | 3.00 | 2.25 | 3.75 | 19 | 1.50 | 0.80 | 1.10 | 2.55 | 2.90 |
| 4 | 6.00 | 1.05 | 2.60 | 8.00 | 3.60 | 20 | 1.45 | 0.80 | 1.10 | 2.40 | 2.80 |
| 5 | 5.50 | 1.00 | 2.40 | 9.20 | 3.40 | 21 | 1.40 | 0.80 | 1.15 | 2.25 | 2.70 |
| 6 | 5.00 | 1.00 | 2.00 | 7.60 | 3.20 | 22 | 1.40 | 3.30 | 1.10 | 2.25 | 2.60 |
| 7 | 3.65 | 1.00 | 1.90 | 5.50 | 3.10 | 23 | 1.30 | 3.00 | 1.50 | 2.20 | 2.40 |
| 8 | 3.20 | 1.05 | 1.50 | 4.30 | 3.10 | 24 | 1.20 | 2.50 | 1.75 | 2.20 | 2.20 |
| 9 | 2.75 | 1.10 | 1.30 | 3.45 | 3.05 | 25 | 3.00 | 2.00 | 1.75 | 2.90 | 2.15 |
| 10 | 2.60 | 1.05 | 1.25 | 2.80 | 3.00 | 26 | 2.00 | 1.70 | 1.70 | 1.90 | 2.10 |
| 11 | 2.20 | 0.95 | 1.25 | 2.00 | 2.90 | 27 | 1.80 | 1.60 | 1.65 | 1.80 | 2.10 |
| 12 | 2.00 | 0.85 | 1.20 | 2.15 | 2.80 | 28 | 1.75 | 1.40 | 1.60 | 2.00 | 2.05 |
| 13 | 1.85 | 0.85 | 1.15 | 6.30 | 2.65 | 29 | 1.60 | 3.60 | 1.50 | 4.00 | 2.00 |
| 14 | 1.70 | 0.90 | 1.15 | 5.00 | 2.50 | 30 | 1.50 | 4.20 | 1.50 | 4.30 | 1.95 |
| 15 | 1.60 | 0.90 | 1.15 | 4.50 | 3.00 | 31 | 1.40 | . . | 1.45 | . . | 1.90 |
| 16 | 1.60 | 0.85 | 1.15 | 3.30 | 3.10 | | | | | | |

[1] In feet.

## DAILY GAUGE-HEIGHT — *Continued*

### C. E. MELTON, *Observer*

| 1897 | | | | | | | | | | | | |
|---|---|---|---|---|---|---|---|---|---|---|---|---|
| | Jan. | Feb. | Mar. | April | May | June | | Jan. | Feb. | Mar. | April | May | June |
| 1 | 1.90 | 3.15 | 3.65 | 4.00 | 3.90 | 2.65 | 17 | 4.05 | 4.60 | 10.90 | 5.30 | 3.10 | 2.50 |
| 2 | 1.90 | 4.40 | 3.60 | 4.00 | 4.00 | 2.70 | 18 | 3.35 | 4.50 | 10.00 | 5.00 | 3.00 | 2.90 |
| 3 | 1.90 | 7.00 | 3.50 | 3.95 | 3.80 | 2.80 | 19 | 3.30 | 4.50 | 9.00 | 4.50 | 3.00 | 2.70 |
| 4 | 1.95 | 7.40 | 3.50 | 4.40 | 3.75 | 2.85 | 20 | 5.40 | 4.35 | 8.50 | 4.20 | 2.90 | 2.60 |
| 5 | 2.00 | 7.10 | 3.60 | 8.50 | 3.65 | 2.90 | 21 | 8.20 | 4.35 | 8.30 | 4.20 | 2.85 | 2.55 |
| 6 | 2.00 | 6.00 | 4.10 | 10.20 | 3.60 | 2.95 | 22 | 7.30 | 4.40 | 8.10 | 4.10 | 2.80 | 2.55 |
| 7 | 2.00 | 6.00 | 10.97 | 11.00 | 3.60 | 2.80 | 23 | 6.50 | 4.85 | 8.00 | 4.00 | 2.75 | 2.50 |
| 8 | 1.95 | 5.20 | 9.30 | 10.50 | 3.60 | 2.70 | 24 | 4.80 | 4.60 | 8.50 | 3.80 | 2.70 | 2.50 |
| 9 | 1.95 | 5.00 | 7.10 | 8.00 | 3.55 | 2.60 | 25 | 3.70 | 4.50 | 7.60 | 3.85 | 2.70 | 2.50 |
| 10 | 1.90 | 4.70 | 5.50 | 7.10 | 3.50 | 2.60 | 26 | 3.50 | 4.00 | 5.00 | 3.70 | 2.70 | 2.40 |
| 11 | 1.90 | 4.90 | 5.30 | 6.50 | 3.50 | 2.65 | 27 | 3.20 | 3.90 | 4.95 | 3.65 | 2.70 | 2.30 |
| 12 | 1.95 | 7.12 | 6.20 | 6.30 | 3.55 | 2.65 | 28 | 3.00 | 3.80 | 4.70 | 3.60 | 2.65 | 2.15 |
| 13 | 1.95 | 6.50 | 10.70 | 6.00 | 3.75 | 2.60 | 29 | 3.00 | . . | 4.50 | 3.60 | 2.65 | 2.00 |
| 14 | 2.10 | 6.10 | 14.10 | 5.80 | 3.60 | 2.50 | 30 | 2.95 | . . | 4.30 | 3.80 | 2.65 | 1.90 |
| 15 | 2.20 | 4.70 | 12.90 | 5.70 | 3.40 | 2.45 | 31 | 3.20 | . . | 4.00 | . . | 2.65 | . . |
| 16 | 4.00 | 4.65 | 11.00 | 5.50 | 3.20 | 2.40 | | | | | | | |

[1] In feet.

## RATING-TABLE

*Drainage Area, 3,300 Square Miles*

| Gauge-height in Feet | Discharge in Cubic Feet per Second | Gauge-height in Feet | Discharge in Cubic Feet per Second | Gauge-height in Feet | Discharge in Cubic Feet per Second | Gauge-height in Feet | Discharge in Cubic Feet per Second |
|---|---|---|---|---|---|---|---|
| . . | . . | 2.00 | 1,890 | 4.00 | 5,830 | 6.00 | 10,550 |
| . . | . . | 2.10 | 2,010 | 4.10 | 6,066 | 6.10 | 10,786 |
| . . | . . | 2.20 | 2,140 | 4.20 | 6,302 | 6.20 | 11,022 |
| . . | . . | 2.30 | 2,280 | 4.30 | 6,538 | 6.30 | 11,258 |
| . . | . . | 2.40 | 2,425 | 4.40 | 6,774 | 6.40 | 11,494 |
| . . | . . | 2.50 | 2,585 | 4.50 | 7,010 | 6.50 | 11,730 |
| . . | . . | 2.60 | 2,760 | 4.60 | 7,246 | 6.60 | 11,966 |
| . . | . . | 2.70 | 2,940 | 4.70 | 7,482 | 6.70 | 12,202 |
| . . | . . | 2.80 | 3,125 | 4.80 | 7,718 | 6.80 | 12,438 |
| . . | . . | 2.90 | 3,310 | 4.90 | 7,954 | 6.90 | 12,674 |
| . . | . . | 3.00 | 3,505 | 5.00 | 8,190 | 7.00 | 12,910 |
| . . | . . | 3.10 | 3,725 | 5.10 | 8,426 | 7.10 | 13,146 |
| 1.20 | 1,060 | 3.20 | 3,950 | 5.20 | 8,762 | 7.20 | 13,382 |
| 1.30 | 1,150 | 3.30 | 4,180 | 5.30 | 8,998 | 7.30 | 13,618 |
| 1.40 | 1,250 | 3.40 | 4,414 | 5.40 | 9,234 | 7.40 | 13,854 |
| 1.50 | 1,350 | 3.50 | 4,650 | 5.50 | 9,470 | 7.50 | 14,090 |
| 1.60 | 1,455 | 3.60 | 4,886 | 5.60 | 9,706 | 7.60 | 14,326 |
| 1.70 | 1,560 | 3.70 | 5,122 | 5.70 | 9,942 | 7.70 | 14,562 |
| 1.80 | 1,665 | 3.80 | 5,358 | 5.80 | 10,178 | 7.80 | 14,798 |
| 1.90 | 1,775 | 3.90 | 5,594 | 5.90 | 10,314 | 7.90 | 15,034 |

## Shallow Ford, near Gainesville, Hall County, Georgia

Two measurements have been made at Shallow Ford, on the Chattahoochee river, four miles from Gainesville, as follows: —

### DISCHARGE MEASUREMENTS

| No. | Date | Measurement Made by | Meter Number | Gauge-height in Feet | Area of Section in Square Feet | Mean Velocity in Feet per Second | Discharge in Cubic Feet per Second |
|---|---|---|---|---|---|---|---|
| 1 | 1896 Mar. 26 | B. M. Hall | 8 | 1.20 | 362 | 2.016 | 730 |
| 2 | Sept. 2 | " " | 8 | 0.40 | 182 | 1.950 | 356 |

On the tributaries of the Chattahoochee river, the following discharge measurements have been made: —

## CHESTATEE RIVER

### Leathers' Ford, Georgia

#### DISCHARGE MEASUREMENT

| No. | Date | Measurement Made by | Meter Number | Gauge-height in Feet | Area of Section in Square Feet | Mean Velocity in Feet per Second | Discharge in Cubic Feet per Second |
|---|---|---|---|---|---|---|---|
| 1 | 1896 Sept. 2 | B. M. Hall | 8 | 0.80 | 102 | 1.372 | 140 |

## PEACHTREE CREEK

### Peachtree Road Bridge, near Atlanta, Georgia

#### DISCHARGE MEASUREMENTS

| No. | Date | Measurement Made by | Meter Number | Gauge-height in Feet | Area of Section in Square Feet | Mean Velocity in Feet per Second | Discharge in Cubic Feet per Second |
|---|---|---|---|---|---|---|---|
| 1 | 1897 May 24 | B. M. Hall | 91 | 0.20 | 35.9 | 1.560 | 56 |
| 2 | June 30 | " " | 14 | 0.00 | 35.5 | 1.135 | 40 |

## SWEETWATER CREEK

### Strickland Bridge, near Austell, Georgia

#### DISCHARGE MEASUREMENTS

| No. | Date | Measurement Made by | Meter Number | Gauge-height in Feet | Area of Section in Square Feet | Mean Velocity in Feet per Second | Discharge in Cubic Feet per Second |
|---|---|---|---|---|---|---|---|
| 1 | 1896 Sept. 4 | B. M. Hall | 8 | .. | 120 | 0.450 | 54.5 |
| 2 | 1897 June 12 | " " | 91 | .. | 138 | 0.666 | 92.0 |

The Southern Railway System, the Western & Atlantic Railroad, the Atlanta & West Point Railroad and the Western Railway of Alabama give easy access to the many fine water-powers of the Apalachicola Basin. The Central Railroad, the Macon & Birmingham Railroad and the Chattanooga, Rome & Columbus Railroad, also, come near a few of these water-powers.

# MOBILE BASIN

## ETOWAH RIVER

### Canton Station, Canton, Georgia

The station at Canton, Cherokee county, was established, as a Geological Survey Station, September 9th, 1896, using the Weather Bureau gauge-rod. It is located at the Cherokee County iron highway bridge, near the railroad depot, in Canton, with Mr. James A. Low, as gauge-observer. The drainage area above this point, is 573 square miles. There is a long record of gauge-heights, for previous years, in the Weather Bureau office. The following is a statement of work done by this Survey: —

#### DISCHARGE MEASUREMENTS

| No. | Date | Measurement Made by | Meter Number | Gauge-height in Feet | Area of Section in Square Feet | Mean Velocity in Feet per Second | Discharge in Cubic Feet per Second |
|---|---|---|---|---|---|---|---|
| 1 | 1896 April 29 | B. M. Hall | 8 | 0.05 | 459 | 2.280 | 590 |
| 2 | July 7 | " " | 8 | 0.59 | 536 | 1.607 | 862 |
| 3 | Sept. 9 | " " | 8 | —0.65 | 390 | 0.560 | 218 |
| 4 | Oct. 28 | " " | 8 | 0.45 | 523 | 1.400 | 733 |
| 5 | " 28 | " " | 8 | 2.25 | 715 | 3.250 | 2,327 |
| 6 | Nov. 27 | " " | 8 | —0.05 | 453 | 0.991 | 449 |
| 7 | 1897 March 17 | " " | 91 | 2.60 | 754 | 3.320 | 2,656 |
| 8 | May 5 | Max Hall | 11 | 0.75 | 541 | 2.336 | 1,264 |
| 9 | June 16 | " " | 11 | 1.27 | 610 | 2.675 | 1,632 |

## DAILY GAUGE-HEIGHT[1]

J. A. Low, *Observer*

| | \multicolumn{4}{c}{1896} | | | | | |
|---|---|---|---|---|---|---|---|---|---|
| | Sept. | Oct. | Nov. | Dec. | | Sept. | Oct. | Nov. | Dec. |
| 1 | . . | 0.00 | 0.00 | 1.00 | 17 | —0.65 | —0.40 | 0.40 | 0.20 |
| 2 | . . | —0.20 | —0.10 | 1.00 | 18 | —0.75 | —0.40 | 0.20 | 0.00 |
| 3 | . . | —0.30 | —0.10 | 0.60 | 19 | —0.75 | —0.40 | 0.00 | 0.00 |
| 4 | . . | —0.30 | 0.00 | 0.30 | 20 | —0.75 | —0.50 | 0.00 | 0.00 |
| 5 | . . | —0.40 | 2.80 | 0.30 | 21 | —0.75 | —0.50 | 0.00 | 0.00 |
| 6 | . . | —0.40 | 0.80 | 0.20 | 22 | —0.60 | —0.50 | 0.00 | 0.00 |
| 7 | . . | —0.50 | 0.60 | 0.10 | 23 | —0.10 | —0.50 | 0.00 | 0.00 |
| 8 | . . | —0.50 | 0.60 | 0.10 | 24 | —0.30 | 0.00 | 0.00 | 0.00 |
| 9 | —0.65 | —0.50 | 0.40 | 0.10 | 25 | —0.40 | —0.10 | 0.00 | 0.00 |
| 10 | —0.60 | —0.30 | 0.40 | 0.10 | 26 | —0.60 | —0.10 | 0.00 | 0.00 |
| 11 | —0.60 | —0.30 | 0.40 | 0.10 | 27 | —0.60 | —0.10 | 0.00 | 0.00 |
| 12 | —0.65 | —0.40 | 0.90 | 0.00 | 28 | —0.60 | 0.00 | 0.00 | —0.10 |
| 13 | —0.70 | —0.20 | 3.60 | 0.00 | 29 | —0.60 | 1.10 | 0.00 | —0.10 |
| 14 | —0.75 | —0.30 | 1.00 | 0.00 | 30 | —0.70 | 1.00 | 1.00 | —0.10 |
| 15 | —0.60 | —0.40 | 0.70 | 0.20 | 31 | . . | 0.00 | . . | —0.10 |
| 16 | —0.55 | —0.40 | 0.70 | 0.40 | | | | | |

[1] In feet.

## DAILY GAUGE-HEIGHT—*Continued* [1]

J. A. Low, *Observer*

| | \multicolumn{12}{c}{1897} |
|---|---|---|---|---|---|---|---|---|---|---|---|---|
| | Jan. | Feb. | Mar. | Apr. | May | June | | Jan. | Feb. | Mar. | Apr. | May | June |
| 1 | —0.10 | 0.60 | 0.80 | 1.60 | 2.00 | 0.10 | 17 | 0.50 | 0.70 | 2.60 | 1.80 | 0.40 | 1.00 |
| 2 | —0.10 | 2.20 | 0.60 | 2.00 | 1.80 | 0.10 | 18 | 2.00 | 0.60 | 2.40 | 1.40 | 0.40 | 0.80 |
| 3 | —0.10 | 1.00 | 0.60 | 2.00 | 1.80 | 0.70 | 19 | 1.60 | 0.60 | 2.00 | 1.40 | 0.40 | 0.60 |
| 4 | —0.10 | 0.80 | 0.60 | 2.60 | 1.60 | 0.50 | 20 | 3.60 | 0.60 | 2.80 | 1.20 | 0.40 | 0.40 |
| 5 | —0.10 | 0.80 | 0.60 | 11.20 | 0.70 | 0.50 | 21 | 3.00 | 0.80 | 2.00 | 1.20 | 0.30 | 0.40 |
| 6 | —0.10 | 0.90 | 3.60 | 5.00 | 0.70 | 0.50 | 22 | 2.00 | 0.80 | 1.00 | 1.00 | 0.30 | 0.30 |
| 7 | —0.10 | 0.80 | 4.00 | 3.00 | 0.70 | 0.40 | 23 | 1.00 | 1.60 | 1.80 | 1.00 | 0.20 | 0.30 |
| 8 | —0.10 | 0.80 | 2.00 | 2.00 | 0.60 | 0.40 | 24 | 0.80 | 1.00 | 1.80 | 1.00 | 0.10 | 0.30 |
| 9 | —0.10 | 0.80 | 1.80 | 3.00 | 0.60 | 0.30 | 25 | 0.70 | 1.00 | 1.60 | 1.00 | 0.10 | 0.30 |
| 10 | —0.10 | 0.80 | 1.80 | 2.60 | 0.50 | 0.30 | 26 | 0.70 | 0.80 | 1.60 | 0.80 | 0.10 | 0.20 |
| 11 | —0.10 | 0.80 | 1.80 | 2.40 | 0.50 | 0.20 | 27 | 0.70 | 0.80 | 1.40 | 0.80 | 0 10 | 0.10 |
| 12 | —0.10 | 0.80 | 2.80 | 2.20 | 0.50 | 0.10 | 28 | 0.60 | 0.80 | 1.20 | 0.80 | 0.10 | 0.10 |
| 13 | —0.10 | 0.80 | 7.20 | 2.00 | 0.50 | 0.10 | 29 | 0.60 | . . | 1.20 | 1.00 | 0.10 | 0.10 |
| 14 | 2.20 | 1.00 | 6.80 | 2.00 | 0.40 | 0.10 | 30 | 0.60 | . . | 1.20 | 1.00 | 0.10 | 0.10 |
| 15 | 1.80 | 0.80 | 4.00 | 2.00 | 0.40 | 0.00 | 31 | 0.60 | . . | 1.20 | . . | 0.10 | . . |
| 16 | 0.90 | 0.80 | 3.60 | 1.80 | 0.40 | 3.00 | | | | | | | |

[1] In feet.

## RATING-TABLE

*Drainage Area, 373 Square Miles*

| Gauge-height in Feet | Discharge in Cubic Feet per Second | Gauge-height in Feet | Discharge in Cubic Feet per Second | Gauge-height in Feet | Discharge in Cubic Feet per Second |
|---|---|---|---|---|---|
| . . . | . . | 1.00 | 1,180 | 3.00 | 3,225 |
| . . . | . . | 1.10 | 1,250 | . . | . . . |
| —0.75 | 200 | 1.20 | 1,340 | . . | . . . |
| —0.70 | 210 | 1.30 | 1,430 | . . | . . . |
| —0.60 | 240 | 1.40 | 1,520 | . . | . . . |
| —0.50 | 270 | 1.50 | 1,610 | . . | . . . |
| —0.40 | 320 | 1.60 | 1,700 | . . | . . . |
| —0.30 | 360 | 1.70 | 1,790 | . . | . . . |
| —0.20 | 410 | 1.80 | 1,880 | . . | . . . |
| —0.10 | 470 | 1.90 | 1,970 | . . | . . . |
| 0.00 | 510 | 2.00 | 2,060 | . . | . . . |
| 0.10 | 565 | 2.10 | 2,160 | . . | . . . |
| 0.20 | 625 | 2.20 | 2,260 | . . | . . . |
| 0.30 | 680 | 2.30 | 2,370 | . . | . . . |
| 0.40 | 750 | 2.40 | 2,480 | . . | . . . |
| 0.50 | 810 | 2.50 | 2,590 | . . | . . . |
| 0.60 | 870 | 2.60 | 2,700 | . . | . . . |
| 0.70 | 950 | 2.70 | 2,830 | . . | . . . |
| 0.80 | 1,025 | 2.80 | 2,960 | . . | . . . |
| 0.90 | 1,110 | 2.90 | 3,100 | . . | . . . |

The minimum discharge, per square mile of drainage area, is 0.35 cubic feet per second.

The other discharge measurements, that have been made, on the Etowah river, are as follows: —

### LADD'S, EAST & WEST R. R., NEAR CARTERSVILLE, GEORGIA

#### DISCHARGE MEASUREMENT

| No. | Date | Measurement Made by | Meter Number | Gauge-height in Feet | Area of Section in Square Feet | Mean Velocity in Feet per Second | Discharge in Cubic Feet per Second |
|---|---|---|---|---|---|---|---|
| 1 | 1896 Aug. 22 | Max Hall | 16 | 0.90 | 317 | 1.40 | 444 |

### ROME,[1] GEORGIA

#### DISCHARGE MEASUREMENTS

| No. | Date | Measurement Made by | Meter Number | Gauge-height in Feet | Area of Section in Square Feet | Mean Velocity in Feet per Second | Discharge in Cubic Feet per Second |
|---|---|---|---|---|---|---|---|
| 1 | 1896 Sept. 24 | Max Hall | 8 | 0.50 | 609 | 1,370 | 834 |
| 2 | 1897 May 1 | " " | 11 | 2.90 | 1,055 | 2,468 | 2,604 |

[1] 2nd Avenue Bridge.

## OOSTANAULA RIVER

RESACA STATION, GORDON COUNTY, GEORGIA

This station was established, as a Geological Survey Station, on July 27th, 1896, using the Weather Bureau gauge-rod. It is located at the Western & Atlantic Railroad bridge at Resaca; and Mr. S. M. Barnett, has been the observer, since the station was established. The drainage area, above this point, is 1,527 square miles.

There is a long record of gauge-heights, for previous years, in the Weather Bureau office. The following is a statement of work done by this Survey at the station: —

### DISCHARGE MEASUREMENTS

*Drainage Area, 1,527 Square Miles*

| No. | Date | Measurement Made by | Meter Number | Gauge-height in Feet | Area of Section in Square Feet | Mean Velocity in Feet per Second | Discharge in Cubic Feet per Second |
|---|---|---|---|---|---|---|---|
| | 1896 | | | | | | |
| 1 | July 27 | Max Hall | 16 | 2.90 | 919 | 1.230 | 1,133 |
| 2 | Aug. 19 | "  " | 16 | 1.47 | 700 | 0.700 | 492 |
| 3 | Oct. 13 | "  " | 11 | 1.70 | 724 | 0.830 | 601 |
| | 1897 | | | | | | |
| 4 | May 25 | Olin P. Hall | 16 | 3.48 | 1,070 | 1.435 | 1,535 |
| 5 | "  29 | "  " | 16 | 3.26 | 998 | 1.392 | 1,389 |
| 6 | June 23 | "  " | 16 | 2.44 | 865 | 1.124 | 972 |

## DAILY GAUGE-HEIGHT [1]

S. M. BARNETT, *Observer*

| | 1896 | | | | | | | | | |
|---|---|---|---|---|---|---|---|---|---|---|
| | Aug. | Sept. | Oct. | Nov. | Dec. | | Aug. | Sept. | Oct. | Nov. | Dec. |
| 1 | 2.10 | 1.20 | 6.50 | 1.50 | 9.25 | 17 | 1.70 | 1.00 | 1.35 | 3.30 | 3.30 |
| 2 | 2.30 | 1.25 | 3.70 | 1.40 | 6.70 | 18 | 1.70 | 1.00 | 1.30 | 3.00 | 3.00 |
| 3 | 2.65 | 1.20 | 2.05 | 1.40 | 4.65 | 19 | 1.50 | 0.95 | 1.20 | 2.70 | 3.00 |
| 4 | 2.40 | 1.25 | 1.65 | 1.35 | 3.90 | 20 | 1.35 | 0.90 | 1.15 | 2.60 | 2.90 |
| 5 | 2.15 | 1.15 | 1.50 | 1.90 | 3.50 | 21 | 1.30 | 0.90 | 1.15 | 2.50 | 2.80 |
| 6 | 2.00 | 1.55 | 1.40 | 3.15 | 3.20 | 22 | 1.25 | 0.85 | 1.20 | 2.35 | 2.60 |
| 7 | 1.90 | 1.30 | 1.30 | 2.10 | 3.00 | 23 | 1.20 | 1.95 | 1.20 | 2.40 | 2.55 |
| 8 | 1.80 | 1.10 | 1.30 | 2.00 | 2.90 | 24 | 1.50 | 1.55 | 1.35 | 2.35 | 2.55 |
| 9 | 1.75 | 1.10 | 1.30 | 2.10 | 3.00 | 25 | 3.20 | 1.25 | 1.90 | 2.25 | 2.40 |
| 10 | 1.75 | 1.05 | 1.20 | 1.85 | 3.50 | 26 | 2.80 | 1.20 | 1.60 | 2.20 | 2.30 |
| 11 | 1.65 | 1.00 | 1.20 | 1.70 | 3.30 | 27 | 1.95 | 1.10 | 1.50 | 2.10 | 2.25 |
| 12 | 1.65 | 1.00 | 1.20 | 3.00 | 3.10 | 28 | 1.70 | 1.10 | 1.40 | 2.15 | 2.20 |
| 13 | 1.75 | 1.50 | 1.55 | 13.65 | 2.90 | 29 | 1.50 | 1.70 | 1.60 | 3.80 | 2.20 |
| 14 | 1.60 | 1.25 | 1.65 | 11.35 | 2.80 | 30 | 1.40 | 8.35 | 1.95 | 8.70 | 2.20 |
| 15 | 1.60 | 1.10 | 1.50 | 11.10 | 4.20 | 31 | 1.30 | . . | 1.55 | . . | 2.20 |
| 16 | 1.70 | 1.05 | 1.45 | 4.25 | 3.90 | | | | | | |

[1] In feet.

## DAILY GAUGE-HEIGHT — *Continued* [1]

### S. M. BARNETT, *Observer*

|     | \multicolumn{6}{c}{1897} | | | | | | | | | | | | |
| --- | Jan. | Feb. | Mar. | Apr. | May | June |     | Jan. | Feb. | Mar. | Apr. | May | June |
| 1   | 2.20 | 3.80  | 4.60  | 7.30  | 5.40 | 4.25 | 17  | 4.10 | 5.40  | 25.30 | 7.00 | 4.70 | 2.70 |
| 2   | 2.20 | 13.90 | 4.40  | 11.30 | 5.05 | 3.50 | 18  | 5.40 | 4.70  | 23.80 | 6.20 | 4.40 | 2.90 |
| 3   | 2.20 | 14.00 | 4.20  | 12.30 | 4.60 | 3.50 | 19  | 5.20 | 4.50  | 21.30 | 5.80 | 4.10 | 2.75 |
| 4   | 2.20 | 13.28 | 4.30  | 12.50 | 4.40 | 3.65 | 20  | 4.40 | 4.50  | 18.90 | 5.50 | 3.95 | 2.60 |
| 5   | 3.00 | 8.70  | 5.90  | 18.50 | 4.25 | 3.35 | 21  | 9.60 | 5.00  | 18.20 | 5.30 | 3.85 | 2.60 |
| 6   | 3.05 | 6.20  | 10.50 | 20.30 | 4.10 | 3.15 | 22  | 8.70 | 4.60  | 18.40 | 5.10 | 3.80 | 2.45 |
| 7   | 2.75 | 7.30  | 18.00 | 19.60 | 4.00 | 3.00 | 23  | 6.10 | 11.40 | 17.50 | 4.90 | 3.75 | 2.35 |
| 8   | 2.50 | 7.60  | 18.80 | 16.30 | 3.95 | 2.90 | 24  | 5.00 | 12.00 | 12.70 | 4.80 | 3.60 | 2.35 |
| 9   | 2.40 | 7.00  | 19.00 | 10.10 | 3.85 | 3.00 | 25  | 4.40 | 10.60 | 8.40  | 4.80 | 3.50 | 2.45 |
| 10  | 2.30 | 6.00  | 16.20 | 10.40 | 3.85 | 3.15 | 26  | 4.00 | 6.70  | 7.60  | 4.70 | 3.35 | 2.45 |
| 11  | 2.25 | 5.80  | 10.70 | 8.60  | 3.95 | 2.90 | 27  | 3.70 | 5.70  | 6.70  | 4.70 | 3.30 | 2.30 |
| 12  | 2.25 | 8.60  | 16.50 | 7.60  | 4.90 | 2.75 | 28  | 3.40 | 5.10  | 6.60  | 4.60 | 3.25 | 2.25 |
| 13  | 2.25 | 9.80  | 21.70 | 6.80  | 5.45 | 2.70 | 29  | 2.70 | . .   | 6.00  | 4.40 | 3.25 | 3.50 |
| 14  | 5.45 | 7.70  | 21.70 | 6.40  | 8.45 | 2.60 | 30  | 2.10 | . .   | 6.00  | 4.30 | 3.15 | 2.90 |
| 15  | 7.50 | 6.40  | 24.60 | 6.80  | 8.75 | 2.55 | 31  | 3.50 | . .   | 6.00  | . .  | 3.50 | . .  |
| 16  | 5.10 | 5.70  | 26.00 | 8.20  | 5.70 | 2.60 |     |      |       |       |      |      |      |

[1] In feet.

## RATING-TABLE

*Drainage Area, 1,527 Square Miles*

| Gauge-height in Feet | Discharge in Cubic Feet per Second | Gauge-height in Feet | Discharge in Cubic Feet per Second | Gauge-height in Feet | Discharge in Cubic Feet per Second | Gauge-height in Feet | Discharge in Cubic Feet per Second |
|---|---|---|---|---|---|---|---|
| 0.85 | 345 | 1.70 | 601 | 2.60 | 995 | 3.50 | 1,547 |
| 0.90 | 355 | 1.80 | 640 | 2.70 | 1,050 | 3.60 | 1,615 |
| 1.00 | 375 | 1.90 | 675 | 2.80 | 1,105 | 3.70 | 1,684 |
| 1.10 | 400 | 2.00 | 715 | 2.90 | 1,162 | 3.80 | 1,755 |
| 1.20 | 427 | 2.10 | 760 | 3.00 | 1,225 | 3.90 | 1,827 |
| 1.30 | 454 | 2.20 | 802 | 3.10 | 1,287 | 4.00 | 1,900 |
| 1.40 | 485 | 2.30 | 850 | 3.20 | 1,350 | | |
| 1.50 | 525 | 2.40 | 898 | 3.30 | 1,414 | | |
| 1.60 | 565 | 2.50 | 948 | 3.40 | 1,480 | | |

The minimum discharge, per square mile of drainage area, is 0.226 cubic feet per second.

As there is a Weather Bureau gauge on the Oostanaula river, at Rome, with a long record of gauge-heights, it has been thought advisable, to make a series of discharge measurements at Rome. But, as the gauge-height, at this point, is not entirely governed by the amount of water, flowing in the stream, being perceptibly affected, by the condition of the Etowah river, which unites with the Oostanaula a short distance below, the following discharge measurement cannot be used to make a rating-table:—

## DISCHARGE MEASUREMENTS

| No. | Date | Measurement Made by | Meter Number | Gauge-height in Feet | Area of Section in Square Feet | Mean Velocity in Feet per Second | Discharge in Cubic Feet per Second |
|---|---|---|---|---|---|---|---|
| 1 | 1896 Sept. 24 | Max Hall (at 5th Ave. bridge) | 8 | 0.20 | 726 | 0.517 | 375 |
| 2 | Oct. 15 | Max Hall (at 5th Ave. bridge) | 11 | 0.35 | 741 | 0.770 | 572 |
| 3 | 1897 May 7 | Max Hall (at 5th Ave. bridge) | 11 | 2.75 | 1,170 | 1.753 | 2,042 |
| 4 | Oct. 15 | Max Hall (at 2nd Ave. bridge) | 11 | 0.35 | 766 | 0.770 | 591 |

## COOSAWATTEE RIVER

### Carter's Station, Carter's, Murray County, Georgia

This station was established August 15th, 1896. It is at the head of navigation; and it has large water-powers immediately above it. Col. S. M. Carter is the observer. The drainage area, above this point, is 532 square miles. The following is a statement of the work done at station: —

### DISCHARGE MEASUREMENTS

| No. | Date | Measurement Made by | Meter Number | Gauge-height in Feet | Area of Section in Square Feet | Mean Velocity in Feet per Second | Discharge in Cubic Feet per Second |
|---|---|---|---|---|---|---|---|
| 1 | 1896 Aug. 15 | Max Hall | 16 | 0.90 | 244 | 1.310 | 320 |
| 2 | "  17 | "  " | 16 | 0.95 | 240 | 1.320 | 319 |
| 3 | Oct. 10 | "  " | 11 | 0.55 | 197 | 1.150 | 228 |
| 4 | 1897 May 22 | "  " | 16 | 2.10 | 379 | 2.150 | 815 |
| 5 | "  24 | "  " | 16 | 1.95 | 369 | 2.089 | 771 |
| 6 | "  26 | Olin Hall | 16 | 1.88 | 352 | 2.020 | 712 |
| 7 | "  28 | "  " | 16 | 1.85 | 346 | 2.017 | 698 |
| 8 | June 1 | "  " | 16 | 1.90 | 358 | 2.020 | 723 |
| 9 | "  15 | "  " | 16 | 1.50 | 312 | 1.745 | 544 |
| 10 | "  28 | "  " | 16 | 1.33 | 290 | 1.634 | 474 |

*APPENDIX* 147

## DAILY GAUGE-HEIGHT [1]

Col. S. M. Carter, *Observer*

|  | 1896 | | | | | 1897 | | | | |
|---|---|---|---|---|---|---|---|---|---|---|
|  | Aug. | Sept. | Oct. | Nov. | Dec. | Jan. | Feb. | Mar. | April | May | June |
| 1 | . . | 0.75 | 1.25 | 0.80 | 2.50 | 1.30 | 1.40 | 1.05 | 4.05 | 4.00 | 1.90 |
| 2 | . . | 0.75 | 1.10 | 0.85 | 2.25 | 1.25 | 4.00 | 1.95 | 4.10 | 3.25 | 1.90 |
| 3 | . . | 0.75 | 1.00 | 1.00 | 2.00 | 1.25 | 3.00 | 1.85 | 5.00 | 3.75 | 1.90 |
| 4 | . . | 0.70 | 0.95 | 1.05 | 2.00 | 1.25 | 2.15 | 1.85 | 9.00 | 2.50 | 2.20 |
| 5 | . . | 0.70 | 0.95 | 3.10 | 1.90 | 1.20 | 2.40 | 1.80 | 15.00 | 2.40 | 2.00 |
| 6 | . . | 0.65 | 0.90 | 1.25 | 1.80 | 1.20 | 2.40 | 9.00 | 4.50 | 2.35 | 1.90 |
| 7 | . . | 0.65 | 0.80 | 1.00 | 1.80 | 1.20 | 2.50 | 5.10 | 4.00 | 2.30 | 1.80 |
| 8 | . . | 0.60 | 0.70 | 1.00 | 1.75 | 1.20 | 2.55 | 4.00 | 3.50 | 2.20 | 1.80 |
| 9 | . . | 0.60 | 0.60 | 0.90 | 1.65 | 1.15 | 2.55 | 3.50 | 3.50 | 2.15 | 1.70 |
| 10 | . . | 0.65 | 0.50 | 0.90 | 1.60 | 1.15 | 2.50 | 3.50 | 5.50 | 2.20 | 1.60 |
| 11 | . . | 0.70 | 0.50 | 6.05 | 1.50 | 1.15 | 2.50 | 3.60 | 5.00 | 2.50 | 1.60 |
| 12 | . . | 0.65 | 0.80 | 3.50 | 1.40 | 1.20 | 2.70 | 19.30 | 4.50 | 2.50 | 1.60 |
| 13 | . . | 0.60 | 0.90 | 2.60 | 1.40 | 4.15 | 2.50 | 11.50 | 4.30 | 2.50 | 1.50 |
| 14 | . . | 0.55 | 0.80 | 1.40 | 1.35 | 2.20 | 2.50 | 11.25 | 4.00 | 2.50 | 1.50 |
| 15 | 0.90 | 0.55 | 0.75 | 1.00 | 2.50 | 2.10 | 2.10 | 10.00 | 3.50 | 2.40 | 1.50 |
| 16 | 0.90 | 0.55 | 0.70 | 0.90 | 2.50 | 2.15 | 2.10 | 8.00 | 4.50 | 2.30 | 2.70 |
| 17 | 0.95 | 0.50 | 0.65 | 0.95 | 2.35 | 2.20 | 2.05 | 5.50 | 3.50 | 2.30 | 1.80 |
| 18 | 0.90 | 0.50 | 0.65 | 0.90 | 2.20 | 2.00 | 2.00 | 5.00 | 3.30 | 2.30 | 1.60 |
| 19 | 0.85 | 0.45 | 0.60 | 0.90 | 2.05 | 2.00 | 2.00 | 6.00 | 3.25 | 2.20 | 1.50 |
| 20 | 0.80 | 0.50 | 0.55 | 0.90 | 2.00 | 2.15 | 2.00 | 6.00 | 3.20 | 2.20 | 1.50 |
| 21 | 0.80 | 0.50 | 0.55 | 0.85 | 1.85 | 4.10 | 2.05 | 5.10 | 3.10 | 2.10 | 1.50 |
| 22 | 0.80 | 0.55 | 0.60 | 0.85 | 1.85 | 2.15 | 2.10 | 5.00 | 3.00 | 2.10 | 1.40 |
| 23 | 0.75 | 0.65 | 0.80 | 0.85 | 1.80 | 2.10 | 7.00 | 4.80 | 3.95 | 2.00 | 1.40 |
| 24 | 0.95 | 0.75 | 1.30 | 0.90 | 1.70 | 2.00 | 3.50 | 4.50 | 3.95 | 2.00 | 1.50 |
| 25 | 0.95 | 0.65 | 0.95 | 1.00 | 1.60 | 2.00 | 2.50 | 4.00 | 3.90 | 1.90 | 1.40 |
| 26 | 0.95 | 0.60 | 0.60 | 1.00 | 1.50 | 1.90 | 2.40 | 3.75 | 3.80 | 1.90 | 1.40 |
| 27 | 0.90 | 0.60 | 0.60 | 0.95 | 1.50 | 1.70 | 2.30 | 3.50 | 3.70 | 1.80 | 1.40 |
| 28 | 0.90 | 0.55 | 0.70 | 1.25 | 1.45 | 1.50 | 2.20 | 3.35 | 3.65 | 1.80 | 1.40 |
| 29 | 0.85 | 1.60 | 1.25 | 1.25 | 1.40 | 1.40 | . . | 3.25 | 3.60 | 1.80 | 2.50 |
| 30 | 0.85 | 1.40 | 0.90 | 3.50 | 1.40 | 1.30 | . . | 3.10 | 3.50 | 2.50 | 1.50 |
| 31 | 0.80 | . . | 0.80 | . . | 1.35 | 1.20 | . . | 3.00 | . . | 2.00 | . . |

[1] In feet.

## RATING-TABLE

*Drainage Area, 532 Square Miles*

| Gauge-height in Feet | Discharge in Cubic Feet per Second | Gauge-height in Feet | Discharge in Cubic Feet per Second | Gauge-height in Feet | Discharge in Cubic Feet per Second | Gauge-height in Feet | Discharge in Cubic Feet per Second | Gauge-height in Feet | Discharge in Cubic Feet per Second |
|---|---|---|---|---|---|---|---|---|---|
| 0.45 | 202 | 0.90 | 337 | 1.40 | 504 | 1.90 | 723 | 2.40 | 973 |
| 0.50 | 215 | 1.00 | 338 | 1.50 | 544 | 2.00 | 771 | 2.50 | 1,026 |
| 0.60 | 242 | 1.10 | 400 | 1.60 | 585 | 2.10 | 820 | | |
| 0.70 | 269 | 1.20 | 433 | 1.70 | 628 | 2.20 | 870 | | |
| 0.80 | 296 | 1.30 | 468 | 1.80 | 675 | 2.30 | 921 | | |

The minimum discharge, per square mile of drainage area, is 0.38 cubic feet per second.

In order to establish the value of the water-powers on the Coosawattee river, above the mouth of Talking Rock creek, a large tributary, which enters the river, about a half mile above Carter's Station, the following measurements have been made on this creek, at its mouth, the gauge-heights, referred to, being those on the river, at Carter's, at the times the measurements were made. The drainage area of Talking Rock creek is 150 square miles.

### DISCHARGE MEASUREMENTS

| No. | Date | Measurement Made by | Meter Number | River Gauge-height in Feet [1] | Area of Section in Square Feet | Mean Velocity in Feet per Second | Discharge in Cubic Feet per Second |
|---|---|---|---|---|---|---|---|
| 1 | 1896 Oct. 10 | Max Hall | 11 | 0.55 | 28 | 1.250 | 35 |
| 2 | 1897 May 24 | " " | 16 | 1.95 | 75 | 1.565 | 117 |
| 3 | June 28 | Olin P. Hall | 16 | 1.33 | 45 | 1.253 | 56 |

[1] At Carter's Station.

A discharge measurement was also made on Salacoa creek, near its mouth, at Nesbitt's bridge, in Gordon county, on June 23d, 1896, when the gauge at Carter's stood at 1.40 feet, and the gauge at Resaca, at 2.35 feet. Measurement made by Olin P. Hall; meter number, 16; area of section, 84 square feet; mean velocity, 0.40 feet per second; discharge, 34 cubic feet per second.

This completes the statement of the work, done on the Mobile Basin, in Georgia.

Very extensive measurements have been made on this basin, in Alabama, on the Coosa and Tallapoosa rivers, whose head-waters come from Georgia.

The railroads, that give access to the water-powers of the Mobile Basin, in Georgia, are the Atlanta, Knoxville & Northern, the Western & Atlantic, The Southern, The Chattanooga, Rome & Columbus, and The East & West.

## TENNESSEE BASIN

There is a regular station of the U. S. Geological Survey at Murphy, N. C., on the Hiawassee river; and discharge measurements have been made on the same river, at Reliance, Tennessee. A great part of this water comes from Georgia; and the measurements will be useful, in the future, for furnishing a water-shed formula, to apply to Georgia streams of the water-shed. These rapid mountain streams in Georgia, which furnish a great part of the waters of the Hiawassee and Ocoee rivers, will be measured, at low-water, during the coming autumn, by this Survey. The only measurement made, so far, on these streams, is at Mineral Bluff, on the Ocoee river (also called the Toccoa river). This measurement was made at extreme low water by the writer, on October 15th, 1896, with meter No. 8. Area of cross-section, 332 square feet; mean velocity, 0.443 feet per second; discharge, 148 cubic feet per second. The Georgia water-powers of the Tennessee Basin are mainly in Fannin, Union and Towns counties, and are reached by the Atlanta, Knoxville & Northern Railroad.

# INDEX

## A

| | |
|---|---|
| Advisory Board | 3 |
| Alcovy River | 16 |
| Almon, Newton County | 122 |
| Altamaha Basin | 14, 41, 112 |
| ————, Important Streams | 41 |
| ————, Utilized Power | 47 |
| ————, Water Powers | 44 |
| Altamaha River | 15 |
| Anthony's Falls | 16 |
| Apalachicola Basin | 14, 28, 123 |
| ————, Important Streams | 28 |
| ————, Utilized Power | 37 |
| ————, Water Powers | 32 |
| Appendix | 103 |
| Atlanta & Florida R. R. | 83 |
| Atlanta & West Point R. R. | 135 |
| Atlanta, Knoxville & Northern R. R. | 149, 150 |
| Atlanta Rainfall | 17, 18 |
| Augusta | 110 |

## B

| | |
|---|---|
| Barnes Shoals | 81 |
| Barnett, S. M. | 104, 141 |
| Barnett's Shoal | 82 |
| Barrow, D. C. | 19 |
| Bass, W. T. | 117 |
| Big Potato Creek | 74 |
| ————, Cross-section | 74 |
| ————, Flow at Nelson's Mill | 74 |
| ————, Fluctuation Table | 75 |
| Brunswick & Western R. R. | 84, 85, 90, 91 |
| Bull Sluice Shoal | 80 |

## C

| | |
|---|---|
| Canton Station | 136 |
| ————, Daily Gauge-heights | 137, 138 |
| ————, Discharge Measurements | 136 |
| ————, Rating-table | 139 |
| Carden Shoal | 81 |
| Carey Station | 112 |
| ————, Daily Gauge-heights | 113 |
| ————, Discharge Measurements | 112 |
| ————, Rating-table | 114 |
| Carlton Station | 111 |
| ————, Discharge Measurements | 111 |
| Carter, S. M. | 104, 146 |
| Carter's Station | 146 |
| ————, Daily Gauge-heights | 147 |
| ————, Discharge Measurements | 146 |
| ————, Rating-table | 148 |
| Cary, J. L. | 104, 112 |
| Cedar Shoals | 82 |
| Central of Georgia R'y | 83, 84, 94, 95, 96, 97, 98, 124, 135 |
| Chattahoochee River | 14, 15, 16, 80, 125 |
| ————, Cross-sections | 64, 65, 68 |
| ————, Flow at Columbus | 68 |
| ————, ———— Roswell Bridge | 64 |
| ————, ———— West Point | 66 |
| ————, Fluctuation Tables | 65, 67, 69 |
| Chattahoochee Water-shed | 17 |
| Chattanooga, Rome & Columbus R. R. | 135, 149 |
| Chestatee River | 134 |
| Cochran Shoal | 81 |
| Columbus Southern R. R. | 86 |
| Commonwealth of Georgia, Henderson's | 19 |
| Coosa River | 14 |
| Coosawattee River | 16, 146 |
| Cotton States and International Exposition | 13 |

## D

| | |
|---|---|
| Dallis, P. A. | 104 |
| Daniels' Mill | 81 |
| Devil's Race-course | 81 |
| Drainage Basins | 14 |
| Dripping Rock Shoal | 81 |
| Dublin Station | 114 |
| ————, Daily Gauge-heights | 115 |
| ————, Discharge Measurements | 114 |

## E

| | |
|---|---|
| East and West R. R. | 149 |
| East Georgia & Florida R. R. | 91 |
| East Tennessee, Va. & Georgia R. R. | 83, 86, 87, 88 |
| Elevations | 191 |
| ———— on Railroad Lines | 83, 86 |
| Etowah River | 16, 136, 140 |

## F

| | |
|---|---|
| Flat Shoals | 81 |
| Flint River | 14, 70, 123 |
| ————, Cross-sections | 70, 72 |
| ————, Flow at M. and B. R. R. Bridge | 72 |
| ————, ———— Sullivan's Mill | 70 |
| ————, Fluctuation Tables | 71, 73 |

# INDEX

## G

Georgia Factory Shoal ..................... 82
——— Midland & Gulf R. R. ......... 83, 86
——— Pacific R. R. ..................... 83, 88, 89
———, Railroad ..................... 111, 116, 122
——— Southern & Florida R. R. ...... 83, 85, 93, 94

## H

Hall, Max, ..................................... 104
———, Olin P. ..................................... 104
Hand-book of Georgia, Jane's, ............... 19
Harper Shoals ..................................... 81
Hiawassee River ..................................... 150
High Falls of the Towaliga ..................... 82
High Shoals ..................................... 82
Holton ..................................... 82
Hurricane Shoals ..................................... 82

## I

Increase in Value of Water-powers, Recent, ... 9
Indian Fishery ..................................... 82
Introduction ..................................... 7
Island Ford Shoal ..................................... 80

## K

Key's Ferry ..................................... 81

## L

Ladd's, East & West R. R. ..................... 140
Lamar's Mill ..................................... 81
Leathers' Ford ..................................... 134
———, Discharge Measurements ............... 134
Letter of Transmittal ..................... 5
Little, Dr. George, ..................... 8
Locke, C. A., ..................................... 19
Low, J. A., ..................................... 104, 139
Lowry, J. H., ..................................... 104, 125
Low-water Measurements, Discussion on, ...... 106

## M

McElroy's Mill ..................................... 82
Macon & Birmingham R. R. ..................... 135
——— Bridge Station ............... 173
Macon & Dublin R. R. ............... 98, 99, 100
——— Station ..................................... 117
———, Daily Gauge-heights..118, 119, 120
———, Discharge Measurements ........... 117
———, Rating-table ..................... 121
Melton, C. E. ..................................... 124, 130
Mercer, J. P., ..................................... 104, 117
Methods and Results of Recent Work ....... 105
Milledgeville ..................................... 116
———, Discharge Measurements ........... 116
Mineral Bluff ..................................... 150
———, Discharge Measurements ........... 150
Mobile Basin ..................................... 14, 139
———, Important Streams ............... 20
———, Utilized Power ............... 26
———, Water-powers ............... 22

Molena ..................................... 123
———, Discharge Measurements ..................... 123
Moore, J. A., ..................................... 104, 123

## N

Newell, F. H., ..................................... 103
Newton Factory ..................................... 82

## O

Oakdale Station ..................................... 125
———, Daily Gauge-heights.126, 127, 128
———, Discharge Measurements.125, 126
———, Rating-table ..................... 129
Ocklockonee Basin ..................................... 15
———, Utilized Power ..................... 58
Ocmulgee River ..................... 14, 16, 76, 117
———, Cross-sections ..................... 76, 78
———, Flow at Juliette ..................... 78
———, Macon ..................... 76
———, Fluctuation Tables ............. 77, 79
Ocmulgee Water-shed ..................................... 16
Ocoee River ..................................... 150
Oconee ..................................... 14, 16, 112
Ogeechee Basin ..................... 15, 52
———, Utilized Power ..................... 53
Ogeechee River ..................................... 15
Oostanaula River ..................................... 141

## P

Peachtree Creek ..................................... 135
——— Road Bridge ..................... 135
———, Discharge Measurements ............. 135
Pittman Ferry ..................................... 81
Pfeiffer, Peter, ..................................... 104, 108
Porter Dale Mills ..................................... 82
Porter Mills ..................................... 62, 63
——— Shoals ..................................... 80
Power, S. P., Jr., ..................................... 111
Price's Shoal ..................................... 82
Princeton Factory ..................................... 82

## R

Redding, R. J., ..................................... 18
Resaca Station ..................................... 141
———, Daily Gauge-heights ....... 142, 143
———, Discharge Measurements ........ 141
———, Rating-table ..................... 144
Reynolds' ..................................... 124
———, Discharge Measurements ........... 124
Rogers' Shoal ..................................... 81
Rome ..................................... 140, 144
———, Discharge Measurements ........ 140, 145
Roswell ..................................... 80
——— Bridge ..................................... 64, 65

## S

Salacoa Creek ..................................... 149
——— Discharge Measurement ........ 149
Satilla and St. Mary's Basin ..................... 15

## INDEX

Savannah Basin .................................. 15, 54
———, Important Streams .............. 54
———, Utilized Power .................. 57
———, Water-powers .................. 56
Savannah, Florida & Western R. R. ....83, 89, 90
Savannah River ............................ 16, 108
Seaboard Air Line ........................ 111, 116
——— R. R. Bridge Station ........ 108
———, Daily Gauge-heights ........... 109
———, Discharge Measurements ...108
———, Rating-table ................... 110
Shallow Ford .................................. 134
———, Discharge Measurements .... 134
Smith's Ferry ................................. 81
Snapping Shoals ............................. 82
Snipes Shoals ................................ 81
Soquee River ................................. 62
———, Cross-section .................... 62
———, Flow at Porter Mills ............ 62
———, Fluctuation Table ............... 63
South River .................................. 16
Southern Fall Line .......................... 15
Southern Railway System..111, 122, 123, 125, 135, 149
Southwestern R. R. ......................... 84
Streams ...................................... 19
———, Flow of, ............................ 59
Strickland Bridge, near Austell .......... 135
———, Discharge Measurement.. 135
Suwannee Basin ............................. 15
———, Utilized Power ................... 58
Suwannee River ............................. 15
Sweetwater Creek .......................... 135

### T

Talking Rock Creek ......................... 148
———, Discharge Measurements.148
Tallapoosa River ............................ 14
Tallassee Bridge Shoal ..................... 82
Tallulah Falls ................................ 16
Tennessee Basin ........................ 20, 150
———, Important Streams ............. 20
Toccoa River ................................ 150
Towaliga ..................................... 16
Tumbling Shoals ............................ 82

### U

U. S. Census Reports ..................... 8, 19
——— Coast and Geodetic Survey .....83, 84, 101
——— Geological Survey ......8, 103, 104, 150
——— Weather Bureau ......8, 17, 18, 104, 115, 136, 141, 144

### V

Vickery's Creek ............................. 80

### W

Western & Atlantic R. R. ....... 92, 135, 141, 149
——— Fall Line ........................... 15
——— Railway of Alabama .............. 135
West Point ............................... 66, 67
——— Station ........................... 130
———, Daily Gauge-heights ..131, 132
———, Discharge Measurements..130
———, Rating-table .................. 133
White and Garner Shoals ................. 82
Wright, Maj. S. B., ........................ 18

### Y

Yellow Jacket Shoals ....................... 81
——— River .......................... 16, 122

www.ingramcontent.com/pod-product-compliance
Lightning Source LLC
Chambersburg PA
CBHW020247170426
43202CB00008B/259